FERRARI
328 • 348 • MONDIAL
Gold Portfolio
1986 -1994

Compiled by
R.M.Clarke

ISBN 1 85520 3278

BROOKLANDS BOOKS LTD.
P.O. BOX 146, COBHAM,
SURREY, KT11 1LG. UK

A-FA86GP

Printed in Hong Kong

BROOKLANDS BOOKS

BROOKLANDS ROAD TEST SERIES

Abarth Gold Portfolio 1950-1971
AC Ace & Aceca 1953-1983
Alfa Romeo Giulietta Gold Portfolio 1954-1965
Alfa Romeo Giulia Berlinas 1962-1976
Alfa Romeo Giulia Coupés 1963-1976
Alfa Romeo Giulia Coupés Gold P. 1963-1976
Alfa Romeo Spider 1966-1990
Alfa Romeo Spider Gold Portfolio 1966-1991
Alfa Romeo Alfasud 1972-1984
Alfa Romeo Alfetta Gold Portfolio 1972-1987
Alfa Romeo Alfetta GTV6 1980-1986
Allard Gold Portfolio 1937-1959
Alvis Gold Portfolio 1919-1967
AMX & Javelin Muscle Portfolio1968-1974
Armstrong Siddeley Gold Portfolio 1945-1960
Aston Martin Gold Portfolio 1972-1985
Aston Martin Gold Portfolio 1985-1995
Audi Quattro Gold Portfolio 1980-1991
Austin A30 & A35 1951-1962
Austin Healey 100 & 100/6 Gold P. 1952-1959
Austin Healey 3000 Gold Portfolio 1959-1967
Austin Healey Sprite 1958-1971
Barracuda Muscle Portfolio 1964-1974
BMW Six Cylinder Coupés 1969-1975
BMW 1600 Collection No.1 1966-1981
BMW 2002 Gold Portfolio1968-1976
BMW 316, 318, 320 (4 cyl.) Gold P. 1975-1990
BMW 320, 323, 325 (6 cyl.) Gold P. 1977-1990
BMW M Series Performance Portfolio1976-1993
BMW 5 Series Gold Portfolio1981-1987
Bricklin Gold Portfolio 1974-1975
Bristol Cars Gold Portfolio 1946-1992
Buick Automobiles 1947-1960
Buick Muscle Cars 1965-1970
Cadillac Automobiles 1949-1959
Cadillac Automobiles 1960-1969
Charger Muscle Portfolio1966-1974
Chevrolet 1955-1957
Chevrolet Impala & SS 1958-1971
Chevrolet Corvair 1959-1969
Chevy II & Nova SS Muscle Portfolio 1962-1974
Chevy El Camino & SS 1959-1987
Chevelle & SS Muscle Portfolio 1964-1972
Chevrolet Muscle Cars 1966-1971
Chevy Blazer 1969-1981
Chevrolet Corvette Gold Portfolio 1953-1962
Chevrolet Corvette Sting Ray Gold P. 1963-1967
Chevrolet Corvette Gold Portfolio 1968-1977
High Performance Corvettes 1983-1989
Camaro Muscle Portfolio 1967-1973
Chevrolet Camaro Z28 & SS 1966-1973
Chevrolet Camaro & Z28 1973-1981
High Performance Camaros 1982-1988
Chrysler 300 Gold Portfolio 1955-1970
Chrysler Valiant 1960-1962
Citroen Traction Avant Gold Portfolio 1934-1957
Citroen 2CV Gold Portfolio 1948-1989
Citroen DS & ID 1955-1975
Citroen DS & ID Gold Portfolio 1955-1975
Citroen SM 1970-1975
Cobras & Replicas 1962-1983
Shelby Cobra Gold Portfolio 1962-1969
Cobras & Cobra Replicas Gold P. 1962-1989
Cunningham Automobiles 1951-1955
Daimler SP250 Sports & V-8 250 Saloon
Gold Portfolio 1959-1969
Datsun Roadsters 1962-1971
Datsun 240Z 1970-1973
Datsun 280Z & ZX 1975-1983
The De Lorean 1977-1993
Dodge Muscle Cars 1967-1970
Dodge Viper on the Road
ERA Gold Portfolio 1934-1994
Excalibur Collection No.1 1952-1981
Facel Vega 1954-1964
Ferrari Dino 1965-1974
Ferrari Dino 308 1974-1979
Ferrari 308 & Mondial 1980-1984
Ferrari 328 • 348 • Mondial Gold Portfolio 1986-1994
Fiat 500 Gold Portfolio 1936-1972
Fiat 600 & 850 Gold Portfolio 1955-1972
Fiat Pininfarina 124 & 2000 Spider 1968-1985
Fiat-Bertone X1/9 1973-1988
Fiat Abarth Performance Portfolio 1972-1987
Ford Consul, Zephyr, Zodiac Mk.I & II 1950-1962
Ford Zephyr, Zodiac, Executive, Mk.III & Mk.IV 1962-1971
Ford Cortina 1600E & GT 1967-1970
High Performance Capris Gold Portfolio 1969-1987
Capri Muscle Portfolio 1974-1987
High Performance Fiestas 1979-1991
High Performance Escorts Mk.I 1968-1974
High Performance Escorts Mk.II 1975-1980
High Performance Escorts 1980-1985
High Performance Escorts 1985-1990
High Performance Sierras & Merkurs
Gold Portfolio 1983-1990
Ford Automobiles 1949-1959
Ford Fairlane 1955-1970
Ford Ranchero 1957-1959
Ford Thunderbird 1955-1957
Ford Thunderbird 1958-1963
Ford Thunderbird 1964-1976
Ford GT40 Gold Portfolio 1964-1987
Ford Bronco 1966-1977
Ford Bronco 1978-1988
Holden 1948-1962
Honda CRX 1983-1987
International Scout Gold Portfolio 1961-1980
Isetta 1953-1964
Iso & Bizzarrini Gold Portfolio 1962-1974
Jaguar and SS Gold Portfolio 1931-1951
Jaguar XK120, 140, 150 Gold P. 1948-1960
Jaguar Mk.VII, VIII, IX, X, 420 Gold P.1950-1970
Jaguar Mk.1 & Mk.2 Gold Portfolio 1959-1969
Jaguar E-Type Gold Portfolio 1961-1971
Jaguar E-Type V-12 1971-1975
Jaguar XJ12, XJ5.3, V12 Gold P. 1972-1990
Jaguar XJ6 Series I & II Gold P. 1968-1979
Jaguar XJ6 Series III 1979-1986
Jaguar XJ6 Gold Portfolio 1986-1994
Jaguar XJS Gold Portfolio 1975-1988
Jaguar XJS Gold Portfolio 1988-1995
Jeep CJ5 & CJ6 1960-1976
Jeep CJ5 & CJ7 1976-1986
Jensen Cars 1946-1967
Jensen Cars 1967-1979
Jensen Interceptor Gold Portfolio 1966-1986
Jensen Healey 1972-1976
Lagonda Gold Portfolio 1919-1964
Lamborghini Countach & Urraco 1974-1980
Lamborghini Countach & Jalpa 1980-1985
Lancia Fulvia Gold Portfolio 1963-1976
Lancia Beta Gold Portfolio 1972-1984
Lancia Delta Gold Portfolio 1979-1994
Lancia Stratos 1972-1985
Land Rover Series I 1948-1958
Land Rover Series II & IIa 1958-1971
Land Rover Series III 1971-1985
Land Rover 90 110 Defender Gold Portfolio 1983-1994
Land Rover Discovery 1989-1994
Lincoln Gold Portfolio 1949-1960
Lincoln Continental 1961-1969
Lincoln Continental 1969-1976
Lotus Sports Racers Gold Portfolio 1953-1965
Lotus & Caterham Seven Gold P. 1957-1989
Lotus Elite 1957-1964
Lotus Elite & Eclat 1974-1982
Lotus Elan Gold Portfolio 1962-1974
Lotus Elan Collection No. 2 1963-1972
Lotus Elan & SE 1989-1992
Lotus Cortina Gold Portfolio 1963-1970
Lotus Europa Gold Portfolio 1966-1975
Lotus Elite & Eclat 1974-1982
Lotus Turbo Esprit 1980-1986
Marcos Cars 1960-1988
Maserati 1965-1970
Maserati 1970-1975
Mercedes 190 & 300 SL 1954-1963
Mercedes 230/250/280SL 1963-1971
Mercedes G Wagen 1981-1994
Mercedes Benz SLs & SLCs Gold P. 1971-1989
Mercedes S & 600 1965-1972
Mercedes S Class 1972-1979
Mercedes SLs Performance Portfolio 1989-1994
Mercury Muscle Cars 1966-1971
Messerschmitt Gold Portfolio1954-1964
MG Gold Portfolio 1929-1939
MG TA & TC Gold Portfolio 1936-1949
MG TD &TF Gold Portfolio 1949-1955
MGA & Twin Cam Gold Portfolio 1955-1962
MG Midget Gold Portfolio1961-1979
MGB Roadsters 1962-1980
MGB MGC & V8 Gold Portfolio 1962-1980
MGB GT 1965-1980
Mini Gold Portfolio 1959-1969
Mini Gold Portfolio 1969-1980
High Performance Minis Gold Portfolio 1960-1973
Mini Cooper Gold Portfolio 1961-1971
Mini Moke Gold Portfolio 1964-1994
Mopar Muscle Cars 1964-1967
Morgan Three-Wheeler Gold Portfolio 1910-1952
Morgan Plus 4 & Four 4 Gold P. 1936-1967
Morgan Cars 1960-1970
Morgan Cars Gold Portfolio 1968-1989
Morris Minor Collection No. 1 1948-1980
Shelby Mustang Muscle Portfolio 1965-1970
High Performance Mustang IIs 1974-1978
High Performance Mustangs 1982-1988
Nash-Austin Metropolitan Gold P. 1954-1962
Oldsmobile Automobiles 1955-1963
Oldsmobile Muscle Cars 1964-1971
Oldsmobile Toronado 1966-1978
Opel GT Gold Portfolio 1968-1973
Packard Gold Portfolio 1946-1958
Pantera Gold Portfolio 1970-1989
Panther Gold Portfolio 1972-1990
Plymouth Muscle Cars 1966-1971
Pontiac Tempest & GTO 1961-1965
Pontiac Muscle Cars 1966-1972
Pontiac Firebird & Trans-Am 1973-1981
High Performance Firebirds 1982-1988
Pontiac Fiero 1984-1988
Porsche 356 Gold Portfolio1953-1965
Porsche 911 1965-1969
Porsche 911 1970-1972
Porsche 911 1973-1977
Porsche 911 Carrera 1973-1977
Porsche 911 Turbo 1975-1984
Porsche 911 SC & Turbo Gold Portfolio 1978-1983
Porsche 911 Carrera & Turbo Gold P. 1984-1989
Porsche 914 Collection No. 1 1969-1983
Porsche 914 Gold Portfolio 1969-1976
Porsche 924 Gold Portfolio 1975-1988
Porsche 928 Performance Portfolio 1977-1994
Porsche 944 Gold Portfolio1981-1991
Range Rover Gold Portfolio 1970-1985
Range Rover Gold Portfolio 1986-1995
Reliant Scimitar 1964-1986
Riley Gold Portfolio 1924-1939
Riley 1.5 & 2.5 Litre Gold Portfolio 1945-1955
Rolls Royce Silver Cloud & Bentley 'S' Series
Gold Portfolio 1955-1965
Rolls Royce Silver Shadow Gold P. 1965-1980
Rolls Royce & Bentley Gold P. 1980-1989
Rover P4 1949-1959
Rover P4 1955-1964
Rover 3 & 3.5 Litre Gold Portfolio 1958-1973
Rover 2000 & 2200 1963-1977
Rover 3500 1968-1977
Rover 3500 & Vitesse 1976-1986
Saab Sonett Collection No.1 1966-1974
Saab Turbo 1976-1983
Studebaker Gold Portfolio 1947-1966
Studebaker Hawks & Larks 1956-1963
Avanti 1962-1990
Sunbeam Tiger & Alpine Gold P. 1959-1967
Toyota MR2 1984-1988
Toyota Land Cruiser 1956-1984
Triumph TR2 & TR3 Gold Portfolio 1952-1961
Triumph TR4, TR5, TR250 1961-1968
Triumph TR6 Gold Portfolio 1969-1976
Triumph TR7 & TR8 Gold Portfolio 1975-1982
Triumph Herald 1959-1971
Triumph Vitesse 1962-1971
Triumph Spitfire Gold Portfolio 1962-1980
Triumph 2000, 2.5, 2500 1963-1977
Triumph GT6 Gold Portfolio 1966-1974
Triumph Stag 1970-1980
TVR Gold Portfolio 1959-1986
TVR Performance Portfolio 1986-1994
VW Beetle Gold Portfolio 1935-1967
VW Beetle Gold Portfolio 1968-1991
VW Beetle Collection No.1 1970-1982
VW Karmann Ghia 1955-1982
VW Bus, Camper, Van 1954-1967
VW Bus, Camper, Van 1968-1979
VW Bus, Camper, Van 1979-1989
VW Scirocco 1974-1981
VW Golf GTI 1976-1986
Volvo PV444 & PV544 1945-1965
Volvo Amazon-120 Gold Portfolio 1956-1970
Volvo 1800 Gold Portfolio 1960-1973
Volvo 140 & 160 Series Gold Portfolio 1966-1975

Forty Years of Selling Volvo

BROOKLANDS ROAD & TRACK SERIES

Road & Track on Alfa Romeo 1949-1963
Road & Track on Alfa Romeo 1964-1970
Road & Track on Alfa Romeo 1971-1976
Road & Track on Alfa Romeo 1977-1989
Road & Track on Aston Martin 1962-1990
R & T on Auburn Cord and Duesenburg 1952-84
Road & Track on Audi & Auto Union 1952-1980
Road & Track on Audi & Auto Union 1980-1986
Road & Track on Austin Healey 1953-1970
Road & Track on BMW Cars 1966-1974
Road & Track on BMW Cars 1975-1978
Road & Track on BMW Cars 1979-1983
R & T on Cobra, Shelby & Ford GT40 1962-1992
Road & Track on Corvette 1953-1967
Road & Track on Corvette 1968-1982
Road & Track on Corvette 1982-1986
Road & Track on Corvette 1986-1990
Road & Track on Datsun Z 1970-1983
Road & Track on Ferrari 1975-1981
Road & Track on Ferrari 1981-1984
Road & Track on Ferrari 1984-1988
Road & Track on Fiat Sports Cars 1968-1987
Road & Track on Jaguar 1950-1960
Road & Track on Jaguar 1961-1968
Road & Track on Jaguar 1968-1974
Road & Track on Jaguar 1974-1982
Road & Track on Jaguar 1983-1989
Road & Track on Lamborghini 1964-1985
Road & Track on Lotus 1972-1981
Road & Track on Maserati 1952-1974
Road & Track on Maserati 1975-1983
R & T on Mazda RX7 & MX5 Miata 1986-1991
Road & Track on Mercedes 1952-1962
Road & Track on Mercedes 1963-1970
Road & Track on Mercedes 1971-1979
Road & Track on Mercedes 1980-1987
Road & Track on MG Sports Cars 1949-1961
Road & Track on MG Sports Cars 1962-1980
Road & Track on Mustang 1964-1977
R & T on Nissan 300-ZX & Turbo 1984-1989
Road & Track on Pontiac 1960-1983
Road & Track on Porsche 1951-1967
Road & Track on Porsche 1968-1971
Road & Track on Porsche 1972-1975
Road & Track on Porsche 1975-1978
Road & Track on Porsche 1979-1982
Road & Track on Porsche 1982-1985
Road & Track on Porsche 1985-1988
R & T on Rolls Royce & Bentley 1950-1965
R & T on Rolls Royce & Bentley 1966-1984
Road & Track on Saab 1972-1992
R & T on Toyota Sports & GT Cars 1966-1984
R & T on Triumph Sports Cars 1953-1967
R & T on Triumph Sports Cars 1967-1974
R & T on Triumph Sports Cars 1974-1982
Road & Track on Volkswagen 1951-1968
Road & Track on Volkswagen 1968-1978
Road & Track on Volkswagen 1978-1985
Road & Track on Volvo 1957-1974
Road & Track on Volvo 1977-1994
R&T - Henry Manney at Large & Abroad
R&T - Peter Egan's "Side Glances"

BROOKLANDS CAR AND DRIVER SERIES

Car and Driver on BMW 1955-1977
Car and Driver on BMW 1977-1985
C and D on Cobra, Shelby & Ford GT40 1963-84
Car and Driver on Corvette 1956-1967
Car and Driver on Corvette 1968-1977
Car and Driver on Corvette 1978-1982
Car and Driver on Corvette 1983-1988
C and D on Datsun Z 1600 & 2000 1966-1984
Car and Driver on Ferrari 1955-1962
Car and Driver on Ferrari 1963-1975
Car and Driver on Ferrari 1976-1983
Car and Driver on Mopar 1956-1967
Car and Driver on Mopar 1968-1975
Car and Driver on Mustang 1964-1972
Car and Driver on Pontiac 1961-1975
Car and Driver on Porsche 1955-1962
Car and Driver on Porsche 1963-1970
Car and Driver on Porsche 1970-1976
Car and Driver on Porsche 1977-1981
Car and Driver on Porsche 1982-1986
Car and Driver on Saab 1956-1985
Car and Driver on Volvo 1955-1986

BROOKLANDS PRACTICAL CLASSICS SERIES

PC on Austin A40 Restoration
PC on Land Rover Restoration
PC on Metalworking in Restoration
PC on Midget/Sprite Restoration
PC on Mini Cooper Restoration
PC on MGB Restoration
PC on Morris Minor Restoration
PC on Sunbeam Rapier Restoration
PC on Triumph Herald/Vitesse
PC on Spitfire Restoration
PC on Beetle Restoration
PC on 1930s Car Restoration

BROOKLANDS HOT ROD 'MUSCLECAR & HI-PO ENGINES' SERIES

Chevy 265 & 283
Chevy 302 & 327
Chevy 348 & 409
Chevy 350 & 400
Chevy 396 & 427
Chevy 454 thru 512
Chrysler Hemi
Chrysler 273, 318, 340 & 360
Chrysler 361, 383, 400, 413, 426, 440
Ford 289, 302, Boss 302 & 351W
Ford 351C & Boss 351
Ford Big Block

BROOKLANDS RESTORATION SERIES

Auto Restoration Tips & Techniques
Basic Bodywork Tips & Techniques
Basic Painting Tips & Techniques
Camaro Restoration Tips & Techniques
Chevrolet High Performance Tips & Techniques
Chevy Engine Swapping Tips & Techniques
Chevy-GMC Pickup Repair
Chrysler Engine Swapping Tips & Techniques
Custom Painting Tips & Techniques
Engine Swapping Tips & Techniques
Ford Pickup Repair
How to Build a Street Rod
Land Rover Restoration Tips & Techniques
MG 'T' Series Restoration Guide
MGA Restoration Guide
Mustang Restoration Tips & Techniques
Performance Tuning - Chevrolets of the '60s
Performance Tuning - Pontiacs of the '60s

BROOKLANDS MILITARY VEHICLES SERIES

Allied Military Vehicles No.1 1942-1945
Allied Military Vehicles No.2 1941-1946
Complete WW2 Military Jeep Manual
Dodge Military Vehicles No.1 1940-1945
Hail To The Jeep
Land Rovers in Military Service
Military & Civilian Amphibians 1940-1990
Off Road Jeeps: Civ. & Mil. 1944-1971
US Military Vehicles 1941-1945
US Army Military Vehicles WW2-TM9-2800
VW Kubelwagen Military Portfolio 1940-1990
WW2 Jeep Military Portfolio 1941-1945
775

BROOKLANDS BOOKS

CONTENTS

Page	Title	Publication	Month	Day	Year
5	Ferrari 328 GTS Road Test	*Car and Driver*	May		1986
10	Ferrari 328 GTB Road Test	*Motor*	June	21	1986
16	Ferrari 328 GTS Road Test	*Road & Track*	May		1986
22	Ferrari Mondial 3.2 QV Road Test	*Autocar*	June	25	1986
29	Brundle's Dream	*Motor*	June	21	1986
30	Rosso Corsa Comparison Test	*Car*	July		1986
36	Master Blaster - Koenig Twin Turbo 328 GTS	*Motor*	Aug	16	1986
40	Ferrari 328 GTS Road Test	*Motor Trend*	Sept		1986
44	Ferrari 328 GTB Road Test	*Motor Sport*	Mar		1987
48	Ferrari Mondial 3.2 Review	*Automobile Magazine*	Oct		1987
54	Ferrari Fabrication	*Automobile Magazine*	Oct		1987
56	Ferrari Mondial 3.2 Road Test	*Motor*	May	7	1988
63	Enzo's New Hit	*Autocar*	June	15	1988
64	Ferraris for Four Comparison Test	*Car*	July		1988
72	Mondial t Road Test	*Road & Track Special*	July		1989
78	Done to a t	*Autocar & Motor*	June	9	1989
84	Ferrari - Mondial t	*Automobile Magazine*	Aug		1989
88	Testarossa Looks and Performance for New 348	*Autocar & Motor*	Sept	20	1989
90	Ferrari 328 GTS Road Test	*Automobile Magazine*	Aug		1989
94	Ferrari, Nein Danke Road Test	*Autosport*	Dec	21	1989
96	Ferrari 348 tb	*Car South Africa*	Feb		1990
103	Ferrari Mondial t Cabriolet Road Test	*Car and Driver*	July		1990
106	NSX vs. 348 ts Comparison Test	*Road & Track Special*	Jan		1991
115	Mister Meets Master Comparison Test	*Modern Motor*	Mar		1990
120	Ferrari Mondial t Cabriolet Road Test	*Road & Track*	Jan		1991
124	Elaboration - Zagato Ferrari 348 tb	*Sports Car International*	July		1991
130	Horse Play - Ferrari 348 ts	*Modern Motor*	Aug		1991
134	A Family Affair	*Autocar & Motor*	June	17	1992
138	348 vs. F40 vs. F1 Comparison Test	*Performance Car*	Aug		1992
144	Ferrari 348 tb Road Test	*Car and Driver*	Mar		1993
148	Feel Good Factor - 348 Spider	*Fast Lane*	Aug		1993
150	Won if by Land, Two if by Sea Comparison Test	*Motor Trend*	Sept		1993
156	The Stuff of Reims - 348 Spider	*Car*	Nov		1993
162	Competizione! - 348 GT	*Sports Car International*	June		1994
168	Ferrari 348 Spider Road Test	*Road & Track Special*	July		1994

ACKNOWLEDGEMENTS

Ferraristi and regular readers of Brooklands Books will know that we already have books in print on the Dino 308s and the early Mondials. With this book, then, we are bringing the story of the small Ferraris a stage further up to date.

As always, we depend on the generosity and co-operation of those who own the copyright material which we reproduce here. On this occasion, we are pleased to express our sincere gratitude to the publishers of *Autocar, Autocar and Motor, Automobile Magazine, Autosport, Car, Car and Driver. Car International, Car South Africa, Fast Lane, Modern Motor, Motor, Motor Sport, Motor Trend, Performance Car, Road & Track* and *Sports Car International.* We are sure that our readers will also recognise how much these publications have contributed to bringing the Ferrari experience to those who have never had the chance to enjoy the cars in the metal.

R M Clarke

The articles in this book tell the story of the cars which used the 3.2-litre and later 3.4-litre editions of Ferrari's twin-overhead-camshaft V8 engine. There were essentially three ranges - the classic GTB/GTS models, their tb/ts replacements, and the controversial four-seater Mondial.

The GTB coupé and GTS targa-top models were beautiful creations by Pininfarina which first appeared in the mid-1970s with a 3-litre dry-sump V8 mounted transversely behind the seats. Re-engined with the bored and stroked 3.2-litre V8 in 1985, these models carried on as the 328GTB and 328GTS until 1988.

Their replacements were the 348tb targa-top cars, which featured longitudinally-mounted V8 engines, now further developed as 3.4-litre types. Time will tell how these cars are viewed in the overall context for Ferrari history, but when new they were often criticised for undistinguished styling and for performance which was not sufficiently ahead of that offered by rival machinery.

Like the GTB/GTS range, the Mondial was a well-established model by the time it was equipped with the 3.2-litre V8 in 1985. First introduced in 1980 with the older 3-litre engine, it never quite had the elegance of the two-seater cars, although in cabriolet form it was an extraordinarily attractive machine. The Mondial retained the transverse 3.2-litre V8 until 1989, when it was re-engineered to take the new longitudinal 3.4-litre engine. These later cars were known as Mondial t models, that additional letter standing somewhat illogically for the transverse gearbox which went with the new engine.

Were the GTB and GTS models among the most attractive Ferraris ever built? In my opinion they were - and an aggregate sales figure of 7,412 examples with the 3.2-litre engine suggests that Maranello's customers thought so too. By contrast, the Mondial sold just 987 copies in 3.2-litre coupé form, plus a smaller number of cabriolets. And what of the 3.4-litre engined cars, for which sales figures are not yet available? Read this fascinating collection of reports and form your own views of the "mainstream" Ferraris built between 1985 and 1994.

James Taylor

Ferrari 328GTS

A new look and more power for an old warrior.

• Until proved otherwise, Ferraris are just cars. From where we sit, they are not dipped in gold and they are not fabricated by God in his personal machine shop. They are not able to leap neomodern skyscrapers in a single bound, and you are not issued a license to laud it over anybody else when you slide behind the wheel of one, or even when you have the financial horsepower actually to buy one. What Ferraris are, generally, is fast, small, red, built in tiny volumes, and expensive.

Nevertheless, you cannot deny that there is a special aura about a Ferrari. It's the same aura often associated with the fabulously wealthy or with astonishingly attractive men and women. We are fascinated by such people. We want to be near them. Deep down inside, though, they are probably in possession of souls as battered as ours. When you cut them, they bleed, and when you strip away their money and get beyond their looks, you're left with something entirely human, and maybe not all that terrific after all.

Ferraris are much the same. When you strip a Ferrari of its lore, work your way through its mystique, and get beyond its illusion of omnipotence, what you're left with is an aggregate of steel and aluminum—substances a lot easier to position on the scales of automotive criticism than freeway flash or plain old car lust.

If you look at the newly introduced 328GTS in these terms, you're looking first and foremost at some pretty old iron. The basic platform has been around since 1974, when it appeared on the European market as the Bertone-designed 208GT4 two-plus-two. It was introduced in the U.S. a year later as the 308GT4. Since then, the changes made to the car have been essentially evolutionary. A more shapely Pininfarina two-seater body was introduced to the American market in 1977. The Mondial models were added—first a two-plus-two coupe and then a two-plus-two convertible. Bosch fuel injection re-

placed Weber carburetion in 1980. Three years later, the 2.9-liter engine was upgraded with four-valve heads. Turbo versions were introduced in Europe, offering more power without being saddled with the tariffs imposed on large-engined cars by the Italian tax laws.

For 1986, there is a new badge in town: Ferrari's volume car for the American market is now designated either 328GTB or 328GTS, depending on whether it's a coupe or a targa. "Volume" in this case means around 900 cars per year. The name may be different, and the changes significant, but the car is by and large the same Ferrari that was introduced in 1974. The 308's double-overhead-cam, four-valve, 2927cc V-8 has been bored and stroked to increase its displacement to 3186cc. In addition, the compression ratio has been bumped up from 8.6 to 9.2:1, intake-valve lift has been increased, and the Digiplex ignition system has been replaced by a slightly more refined Microplex system. Other improvements to the engine include redesigned intake runners, larger oil coolers and lines, a different throttle body, slightly redesigned combustion chambers, and different spark plugs. As a result of these changes, the power output jumps from 230 hp at 6800 rpm to 260 at 7000. Torque is also increased, from 190 pounds-feet at 5500 rpm to 213 at the same engine speed.

A number of other mechanical improvements have been sprinkled throughout the 328, but they're not anything that would force Porsche, Nissan, or even Honda to initiate a crash catch-up program. The steering rack is borrowed from the GTO, the front wheel bearings are borrowed from the Mondial, and the shock absorbers, the springs, and the knuckles have been revised. We tried to ascertain to what extent the spring rates and damping values have changed, but Ferrari prefers to play such details close to its vest.

Other changes are apparent to the naked eye. The interior gets a new dashboard, new seats, better climate-control levers and dials, and new door pulls. Exterior cosmetic changes include integrated, body-colored bumpers, redesigned radiator-air intakes, slightly modified wheels, and new door handles.

By far the most significant improvement is the horsepower increase. The last 308 we tested ran to 60 mph in 7.4 seconds and did the quarter-mile in 15.2 seconds at 92 mph. The new 328 scoots to 60 in 5.6 seconds and trips the quarter-mile clock in 14.2 seconds at 97 mph. Together with new, more aerodynamic body panels, the power increase yields a top speed of 153 mph, a solid 9-mph improvement.

If you look at the comparison graphs at

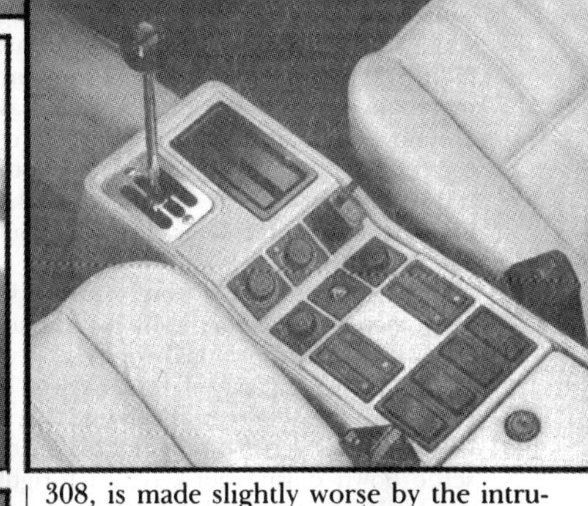

the end of this story, you'll notice that there isn't much that can chase the 328 on the on-ramps. About the only car that can put the hurt on it is the Porsche 911 Turbo. The Porsche does cost some $13,000 less, but stable fees for prancing ponies have never been cheap.

The 328's power delivery is deceptive, which is typical of four-valve engines. You don't get the rush and sudden kick of an engine coming on the cam. What you do get is steady forward propulsion accompanied by the mechano-music of 32 valves, four camshafts, and an exhaust note that's as arrogant as a Neapolitan street punk with a switch-blade.

As a superslab warrior, the 328 has more or less the same strengths and weaknesses as the 308. The old car's resilient suspension and compliant ride motions have been preserved. The new seats, though a little narrow, are very comfortable, and they have enough travel to accommodate Tom Selleck. Rear visibility, limited in the 308, is made slightly worse by the intrusion of the high-mounted brake light in the bottom center of the glass. The rear glass also picks up a lot of glare, so lane changes are best executed after a look over your shoulder.

The shifter is typically Ferrari: stiff and uncooperative when cold, but slicker than a cocaine dealer's lawyer when the linkage is properly adjusted and the transmission is all warmed up. The 328 is one of a handful of cars in the world that you shift even when you don't really need to. You find yourself going down through the gears for every stop sign and traffic light just to hear the motor sing and feel the gears notch into place.

On smooth roads, at legal freeway speeds and beyond, the 328's directional stability and on-center feel are terrific. Unfortunately, even small bumps seem to upset the car slightly, making it wander around in its lane. This quirk is nothing major at legal speeds—you let the car walk a few feet laterally, then nudge it back to the desired trajectory—but it is amplified at higher velocities.

As single-minded and seemingly invincible as it might be as a straight-line racer, the 328 develops a Jekyll-and-Hyde personality in the corners. On a very smooth, tight road at less than ten-tenths, it exhibits decent if not sterling manners, characterized by initial understeer and a gradual transition to a neutral attitude. This behavior isn't exactly precise, but at least it's predictable. Standard operating procedure for entering a corner in the 328 is to dial in

some steering and hold a steady heading while you wait for the front grip to start fading. When it does, you dial in some more steering and sort of pick your way around. Although this technique requires careful attention, it doesn't set off your self-defense alarm bells or send you to the phone for reinforcements. It's work, but it's pleasant work.

In more aggressive driving, your workload increases geometrically with road speed. In fast transitions, you're always aware that the majority of the car's mass is back there over the rear wheels and that it takes that mass a certain amount of time to settle down during quick right-left-right combinations. You really become aware of the mass if you decide to cancel some of the inherent understeer with lift-throttle oversteer. Lifting off for too long will produce more oversteer than you want, and the result is guaranteed to light up every neuron in your brain. It takes a brave and judicious right foot to get the balance just right. We wouldn't call the 328 as tricky as a first-generation 930 Turbo, but the trailing-throttle excitement is there and must occasionally be dealt with.

On bumpy roads, the 328GTS really loses its cool. This is where the vicious Mr. Hyde throws on his cloak and goes looking for dates. Bumps and road irregularities want to take the steering wheel right out of your hands. The cowl shakes, the chassis flexes, and whatever precision was designed into the suspension gets air-mailed back to Maranello with no return address. When the targa top is removed, the lack of rigidity reaches nearly critical levels and the car becomes truly unpleasant to drive.

Ultimately, you walk away from the 328 with the feeling that all that's standing in the way of its becoming a ground-stomping, apex-eating Porsche slayer is maybe six months of development work focused on stiffening its chassis. It already has one of the world's best V-8s, inspired looks, a fine suspension design, and a lock on about two-thirds of the world's automotive magic. If the factory could somehow reinforce its charms by welding some steel stampings into the right places, we would be the last to deny the 328 its proper position near the top of the hierarchy.

In its present form, however, Ferrari's volume car is a disappointment. Its performance envelope has been stretched by the bigger engine, but its flexible chassis is literally an underlying weakness that's just waiting for you to take a ride down the wrong road. What the 328GTS promises is that it can not only run in the thick of the supercar pack but even embarrass 99 percent of the cars in its league. It will deliver on those promises, but only if you stay away from that wrong road.

—Tony Assenza

COUNTERPOINT

• As one who believes that the heart of any Ferrari is its engine, I've always been disappointed by the eight-cylinder sports cars. The different 2.9-liter powerplants have had exotic specifications, but they have never motivated the various GTBs, GTSs, and Mondials with much energy. And knowing that these Ferraris couldn't blow off a good-running Mustang didn't make me very tolerant of their numerous ergonomic and practical shortcomings.

The new 328GTS has turned my opinion around completely. The revamped engine is powerful and willing throughout its rpm range, transforming the GTS into one of the quickest cars on the market. Combined with an improved interior, the new engine makes driving the GTS a genuine pleasure. Even the car's sensuous styling—always its strongest point—has been improved, thanks to the new Boxer-like front and rear ends. With these changes, the GTS deserves to wear its prancing horse as proudly as the best of the twelve-cylinder cars. *—Csaba Csere*

Not long ago, Ferraris were fine, expensive toys that most folks could stand to drive only once or twice a week. The factory has now taken steps to change that dismal state of affairs, making the 328GTS a worthwhile daily-transportation tool. The seats are comfortable; the heating-and-ventilating system works; the exhaust note is mellow and nice. And my favorite Ferrari dealer, Rick Mancuso, of Lake Forest, Illinois, informs me that these Italian fantasies are no longer garage bunnies. His customers pay their ransoms, drive their new cars out the door, and become total strangers. After a brief tour of his facilities, I can confirm that the number of red automobiles leaking vital fluids on the shop floor is unusually low.

I can also tell you that driving the new 328 is a blast. The steering has been relieved of friction. The engine feels torquey at low rpm and appropriately wild in the streets near the redline. The cockpit is full of modern switches, LED monitors, and a number of details transferred from the GTO. The low-mileage demonstrator I drove had a reluctant shifter, but I chalked that up to the price of entry: Ferraris will probably never be so refined that any idiot can hop in and make them sing.

—Don Sherman

There's no doubt about it: something's shaking at the home of the prancing horse. Ferrari has awakened from a deep sleep and now seems to realize that it is in the business of building *cars,* not rolling artwork.

Who would have thought that a silly little 259cc increase in engine displacement would have made the GTS's heart beat with so much more fire? The 328's urgent exhaust note pulsates unlike that of any V-8 you've ever heard. Even plugging along in town is a course in music appreciation.

The signs of change are everywhere, from the redesigned switch gear to the reduction in steering effort. A person could live with this car.

According to the Ferrari dealer who lent us one of his 328s for a test jaunt, quality has become a corporate priority. After all, who wants a 60-grand car with door locks that jam and electrics that go poof?

Paying attention to such mundane matters indicates that Ferrari is once again living in the real world. To that I say, "*Bene, Bene, Bene.*" *—Rich Ceppos*

FERRARI 328GTS

CURRENT BASE PRICE dollars x 1000

CHEVROLET CORVETTE CONVERTIBLE

PORSCHE 911 TURBO

LOTUS ESPRIT TURBO

FERRARI 328GTS

0 15 30 45 60 75

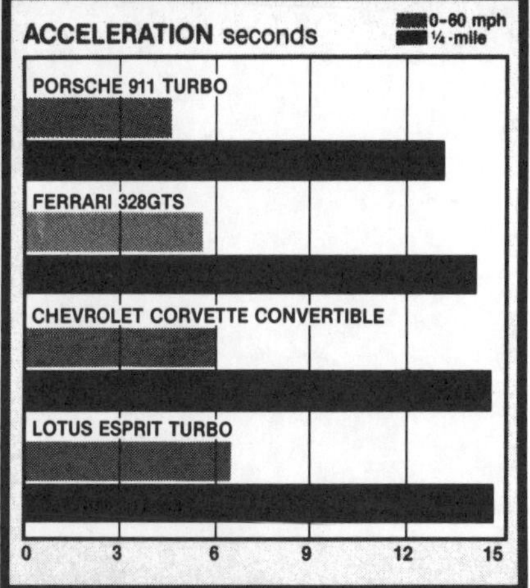

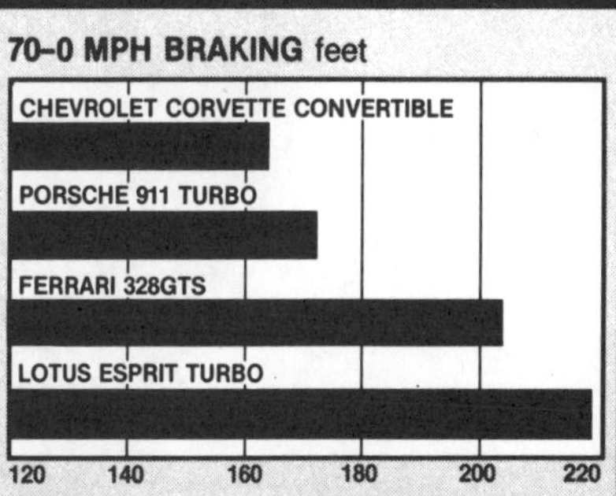

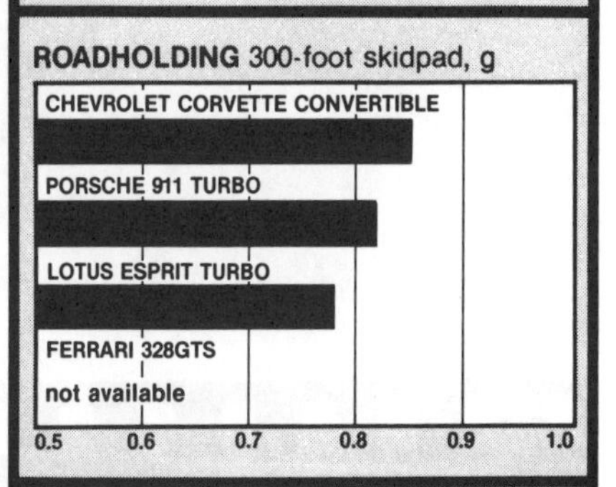

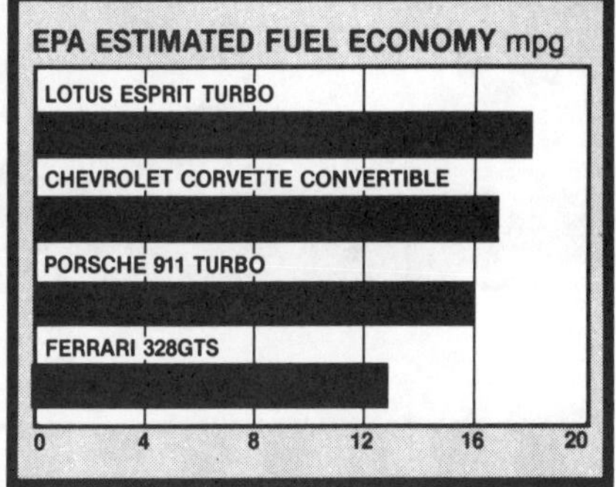

Vehicle type: mid-engine, rear-wheel-drive, 2-passenger, 2-door targa

Price as tested: $64,393

Options on test car: base Ferrari 328GTS, $61,000; Sony sound system, $1023; gas-guzzler tax, $1500; freight, $870

Standard accessories: power windows and locks, A/C, rear defroster

Sound system: Sony CDX-R7 AM/FM-stereo radio/compact disc, 4 speakers

ENGINE

Type .... V-8, aluminum block and heads
Bore x stroke .... 3.27 x 2.90 in, 83.0 x 73.6mm
Displacement .... 194 cu in, 3186cc
Compression ratio 9.2:1
Fuel system .... Bosch K-Jetronic fuel injection
Emissions controls ..... 3-way catalytic converter, feedback fuel-air-ratio control, auxiliary air pump
Valve gear .... belt-driven double overhead cams, 4 valves per cylinder
Power (SAE net) .... 260 bhp @ 7000 rpm
Torque (SAE net) .... 213 lb-ft @ 5500 rpm
Redline .... 7700 rpm

DRIVETRAIN

Transmission 5-speed
Primary-drive ratio 1.07:1
Final-drive ratio .... 4.06:1, limited slip

Gear	Ratio	Mph/1000 rpm	Max. test speed
I	3.08	5.3	41 mph (7700 rpm)
II	2.12	7.8	60 mph (7700 rpm)
III	1.52	10.8	83 mph (7700 rpm)
IV	1.12	14.7	113 mph (7700 rpm)
V	0.83	19.9	153 mph (7700 rpm)

DIMENSIONS AND CAPACITIES

Wheelbase .... 92.5 in
Track, F/R .... 58.1/57.7 in
Length .... 169.1 in
Width .... 67.7 in
Height .... 44.1 in
Ground clearance .... 4.6 in
Curb weight .... 3090 lb
Weight distribution, F/R 41.7/58.3%
Fuel capacity .... 19.6 gal
Oil capacity .... 10.6 qt
Water capacity .... 19.0 qt

CHASSIS/BODY

Type .... unit construction
Body material .... welded steel stampings, aluminum stampings, fiberglass-reinforced plastic

INTERIOR

SAE volume, front seat .... 48 cu ft
trunk space .... 5 cu ft
Front seats .... bucket
Seat adjustments .... fore and aft, seatback angle
General comfort .... poor fair **good** excellent
Fore-and-aft support .... poor fair **good** excellent
Lateral support .... poor fair **good** excellent

SUSPENSION

F: .... ind, unequal-length control arms, coil springs, anti-roll bar
R: .... ind, unequal-length control arms, coil springs, anti-roll bar

STEERING

Type .... rack-and-pinion
Turns lock-to-lock 3.2
Turning circle curb-to-curb .... 39.4 ft

BRAKES

F: .... 11.1 x 0.9-in vented disc
R: .... 11.0 x 0.9-in vented disc
Power assist .... vacuum

WHEELS AND TIRES

Wheel size .... F: 7.0 x 16 in; R: 8.0 x 16 in
Wheel type .... cast aluminum
Tires .... Goodyear NCT VR55/50, F: 205/55VR-16; R: 225/50VR-16
Test inflation pressures, F/R .... 31/34 psi

CAR AND DRIVER TEST RESULTS

ACCELERATION Seconds

Zero to 30 mph 1.9
40 mph 2.9
50 mph 4.3
60 mph 5.6
70 mph 7.4
80 mph 9.4
90 mph 12.3
100 mph 15.1
110 mph 18.7
Top-gear passing time, 30–50 mph 7.5
50–70 mph 7.4
Standing ¼-mile .... 14.2 sec @ 97 mph
Top speed .... 153 mph

BRAKING

70–0 mph @ impending lockup .... 204 ft
Modulation .... poor fair good **excellent**
Fade .... **none** moderate heavy
Front-rear balance .... poor **fair** good

COAST-DOWN MEASUREMENTS

Road horsepower @ 30 mph .... 6 hp
50 mph .... 15 hp
70 mph .... 31 hp

FUEL ECONOMY

EPA city driving .... **13 mpg**
EPA highway driving .... 18 mpg
C/D observed fuel economy .... **16 mpg**

INTERIOR SOUND LEVEL

Idle .... 71 dBA
Full-throttle acceleration .... 94 dBA
70-mph cruising .... 81 dBA
70-mph coasting .... 77 dBA

KODAK EN 100 5039
KODAK EN 100 5039
C393 PPB
31
31A
32
32A
33
33A

FERRARI 328 GTB

Maranello's entry-level exotic is more expensive than all but a handful of car makers' flagships. It's also more beautiful, more exciting and more desirable

C393 PPB
328
GTB

This time Ferrari have stolen the initiative. Rather than wait for arch-rivals Porsche to hike the power of their redoubtable 911 Carrera and erase the 308 QV's small performance advantage altogether, Maranello's engineers have stepped in with the 328, the automotive equivalent of a knee-wobbling left hook to the Teuton's jaw in a needle bout that has been finely poised for the best part of a decade and looks set to keep us guessing for a few years yet.

Ferrari's claim to the Junior Supercar title is more convincing than ever with the 328 which, in addition to its extra swept volume and power, gets a subtle styling update to improve drag and incorporate new Testarossa-style moulded bumpers with integrated side lights and indicators. A deeper chin spoiler lends a more aggressive appearance to the front. At the back a plastic apron now extends down over the unsightly plumbing of the four-pipe exhaust system while, attached to the trailing edge of the roof, is a neat and unobtrusive "ski blade" style wing. The bluffer nose and deeper rump make the 328 – both in GTB guise as tested and as the targa-topped GTS – look slightly stubbier than the 308, but if the sublime purity of the original shape has been corrupted, its striking elegance has not. In our book, this is still the most beautiful of all contemporary exotics – a gorgeous-looking car.

The 328 GTB buyer's £34,750 pays for other improvements, too: a mildly re-worked interior with better dials, switchgear and heating and more attractively contoured alloy wheels, up in size from 6.5 x 15.3 in all round to 7 x 16 in at the front and 8 x 16 in at the rear (and wearing 205/55 and 225/50 Goodyear NCTs.

Otherwise the chassis is much as before with suspension by double wishbones, coil springs and an anti-roll bar front and rear, steering by unassisted rack and pinion and braking by large servo-assisted ventilated discs all round. All this is hung on the traditional tubular steel chassis clothed with those sensuously flowing steel and aluminium body panels.

The chassis cradles Ferrari's 90 degree all-alloy V8 transversely ahead of the driven rear wheels: a limited slip differential is enclosed in unit with the five-speed gearbox mounted behind the engine. By increasing the bore and stroke from 81/71 mm to 83/73.6 mm, the quad-cam 32-valver's capacity goes up from 2927 to 3185 cc and on a higher compression ratio (9.8 instead of 8.8:1) develops 270 bhp at 7000 rpm and 224 lb ft of torque at 5000 rpm, increases of 12.5 and 16.6 per cent respectively. Improved piston design with more pronounced "squish" and better combustion along with a switch to Marelli's most sophisticated Microplex engine management system and smaller (12 mm) spark plugs for better heat dissipation help add muscle, too.

Centre console now displays a pleasing lack of chrome and electronic heater controls work well. Cabin is comfortable but short on headroom

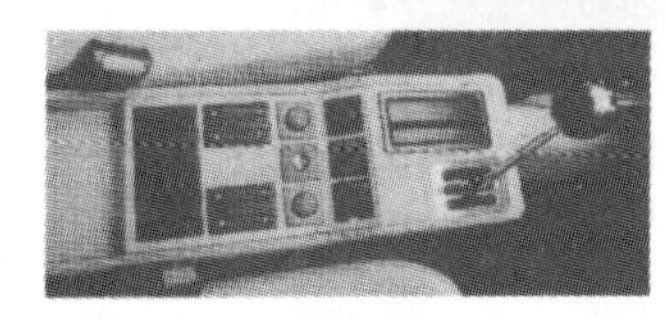

It means that in terms of sheer power, the Ferrari now has significantly more at its disposal than either of its major price/performance rivals, the Porsche 911 Carrera (231 bhp) and the Lamborghini Jalpa (250 bhp), enough to win the power/weight battle in both cases even though the Porsche is nearly 200 kg lighter than the Ferrari. It's also worth remembering that the 911 can put the superb traction afforded by its rear-slung flat-six to fine use off the line.

So although the 328 turns in a stunning 5.5 sec for the sprint from rest to 60 mph, shaving 0.2 sec off the 308 QV's time, the Carrera is quicker still with 5.3 sec. The Lamborghini, a rapid car by any standards, is left trailing with 5.8 sec. To the Porsche's credit, it keeps its nose ahead of the Ferrari's all the way to 100 mph which it reaches in 13.6 sec, the 328 just 0.2 sec behind but the Jalpa well out of the running with 16.0 sec. From here on, however, it's the Ferrari all the way, its more slippery shape and superior top end power stretching an impressive gap by 120 mph and, ultimately, logging a top speed round Millbrook's high-speed bowl of 158.5 mph with a fastest leg of 161.2 mph. Not only does this put the Carrera (151.1 mph) and the Jalpa (147.6 mph) firmly in the shade, it's the fastest speed we've ever recorded round Millbrook, the previous record being held, perhaps unsurprisingly, by the 308 QV with 154.5 mph. Such speeds are very close to Millbrook's "limit" for cars on road tyres, up to 5 mph being lost through scrub. Make no mistake, the 328 is a genuine 160 mph car.

If all that seems too good to be true on just 3 litres, the 328's overtaking punch in fourth and fifth gears is, if anything, even more impressive. Although the 32-valver feels a cammy unit on the road, its measured flexibility is astonishing. As if the 308 QV's ability to put away all the 20 mph increments between 20 and 110 mph in around 5 seconds apiece in fourth gear wasn't enough, the 328 cuts the time taken to cover each increment to around 4.5 sec. Even in fifth, 80-100 mph takes a mere 7.9 sec. Here the Carrera is shouldered out of contention. Without the width of powerband to run the Ferrari's short overall gearing (24.3 against just 21.0 mph/1000 rpm) it never gets on terms, leaving it to the larger-engined but still high-revving Jalpa to challenge the 328 in the lower speed ranges.

Another shock. The Ferrari sounds like a Fiat X1/9, only louder. At least it does at low revs with light throttle openings. The hollow, slightly harsh thrum tends to take admiring onlookers by surprise, so incongruous is its emanation from such an exotic presence. The driver's only consolation is that it doesn't sound quite so boring from the inside and that, around town, fine tractability and a complete absence of temperament do nothing to detract from the "cooking car" illusion.

But illusion it is. Driven with purpose, the 328 has, perhaps, the most exciting engine in production. At around 3000 revs, the Fiat-like hum hardens and acquires a tingling, metallic edge. Beyond 5000, energy and pace build with the exponential fury of a nuclear reaction, the engine howling its savage approval. In the lower gears, the revs will slam into the 7700 rpm limiter with the force of a runaway express unless a watchful eye is trained on the rev counter. Punctuated by well-timed gearchanges, however, the fast-driving experience has race car-realism and is guaranteed to set the adrenalin flowing as urgently as the car itself. The engine should carry a government health warning, and not just because it's so loud.

Drive hard, and petrol is consumed with similar vigour. Our overall consumption of 18.9 mpg is certainly better than the Jalpa's 15.6 mpg but falls some way short of the 911 Carrera's 21.1 mpg and the old 308 QV's 19.6 mpg. A projected touring consumption of 22.5 mpg shows what can be achieved with appropriate restraint and permits a useful maximum range of 367 miles on a 74-litre (16.3-gallon) tankful of 4-star.

The Ferrari's metal gate gearchange is regarded as a challenge by some and a curse by others. In truth it isn't nearly as intimidating to use as its uncompromising looks would suggest and, guided with deliberation, the slim chrome lever slips home with a nuggety precision you come to admire. A more casual approach, however, is punished by obstinate baulking, especially into the dog-leg first. It pays to play the gearbox's tune in the long run: driving pleasure is easily tarnished otherwise. Only between fourth and fifth do the ratios feel anything other than closely stacked but with such a wide power band it isn't a problem. The clutch is surprisingly light and extremely well-cushioned.

It almost goes without saying that impeccable road manners

Engine's architecture is a joy to behold. Noise and power are formidable

are expected of the 328, and in normal fast driving on dry roads you'd have to look hard to find any better. The steering combines lightness with direct gearing and excellent feel. The elegant leather-rimmed wheel doesn't writhe in the driver's hands 911-fashion but there is just enough kickback to let him know when the front wheels are being deflected by bumps.

Helm responses are deliciously quick and accurate without any hint of twitchiness and, in the dry, understeer is the predominant characteristic – strong on tight bends entered too fast, very mild on fast sweepers. Such is the grip generated by the fat Goodyears that even drivers of modest skill can travel almost absurdly quickly on demanding country roads without putting their car control to the test. More ambitious pilots, however, had better know what they're about because, while the 328 is forgiving up to a point, liberal applications of the ample torque will push the tail out a long way even on 70 mph third gear bends, and unless the correct amount of opposite lock is applied quickly, the results can be untidy to say the least. The Ferrari is more slideable than, say, a Lotus Esprit Turbo and, treated with respect, is more fun. But, at the end of the day, it doesn't have as much grip as the Lotus and needs a better man at the wheel to drive near the limit, especially in the wet – though, by the same token, it needs a better man still to master a 911.

The Ferrari's ride quality is seldom less than impressive with more than reasonable small bump absorption around town, an ability to soak up major surface disturbances without structural shuddering and simply superb suspension control at speed. No car in our experience has felt as smooth and composed round Millbrook's far-from-flawless high-speed bowl at 150 mph-plus as the 328. But for their propensity to lock a front wheel prematurely in the wet, the brakes would be beyond criticism, too.

So noisy is the Ferrari's engine that the generally low levels of wind noise, road roar and tyre bump-thump don't count for much. Cruising at 120 mph is tolerable but to go any faster is to drown out the radio or any chance of conversation with your passenger altogether.

For the rest, the 328 differs little from its 308 QV forebear. It remains one of the most practical and habitable of supercars with a pleasantly un-Italianate driving position limited more by a paucity of headroom than legroom. There's a usefully square-shaped luggage compartment behind the engine and an assortment of cabin cubbies for odd-ments stowage. Visibility is excellent by mid-engined supercar standards and both the new instruments and heater/air conditioning controls are a big improvement on the 308's.

Anyone who has visited Ferrari's Maranello factory – and many Ferrari owners regard it as an essential pilgrimage – will have seen the great care with which prancing horse products are assembled, yet detail execution still leaves something to be desired. The Maranello Concessionaires demonstrator we borrowed had wonky door handles and mediocre paint finish in places. That said, the cabin looked convincingly expensive and the whole car felt tight and rattle-free.

Ferrari must still give best to Porsche when it comes to build quality but, in every other respect, the contest between the two big names in junior-league exotica is hotter than ever. The £36,577 Lamborghini Jalpa joins the £34,750 Ferrari and £36,676 Carrera Sport on price and driver appeal but is beginning to look a little slow in this company. Lotus' value-for-money Esprit Turbo (£23,440) is worth a mention, too, and as well as having competitive performance, can out-corner any of its German or Italian rivals.

But ultimately, it is exactly what the Lotus and, to a lesser extent the Porsche, lack that makes the Ferrari great: a charismatic presence.

The 328 is a rare and beautiful car, as close to a work of art as any modern car can be. That it is also faster than ever and easier to live with makes it a car you ache to own. It is, after all, a Ferrari.

PERFORMANCE

WEATHER CONDITIONS

Wind	5-12 mph
Temperature	59 deg F/ 12 deg C
Barometer	29.5 in Hg/998 mbar
Surface	Dry tarmacadam

MAXIMUM SPEEDS

	mph	kph
Banked Circuit (5th)	158.5	255.0
Best ¼ mile	161.2	259.4
Terminal speeds:		
at ¼ mile	103	166
at kilometre	126	203
Speeds in gears (at 7700 rpm):		
1st	43	69
2nd	63	101
3rd	88	142
4th	119	191

ACCELERATION FROM REST

mph	sec	kph	sec
0-30	2.2	0-40	1.9
0-40	3.0	0-60	2.8
0-50	4.3	0-80	4.3
0-60	5.5	0-100	6.0
0-70	7.2	0-120	8.0
0-80	8.9	0-140	10.6
0-90	11.3	0-160	13.7
0-100	13.8	0-180	17.7
0-110	16.7		
Stand'g ¼	14.1	Stdg km	25.4

ACCELERATION IN TOP

mph	sec	kph	sec
20-40	7.4	40-60	4.7
30-50	7.0	60-80	4.3
40-60	7.0	80-100	4.3
50-70	7.0	100-120	4.4
60-80	7.1	120-140	4.5
70-90	7.2	140-160	5.0
80-100	7.9		

ACCELERATION IN 4TH

mph	sec	kph	sec
20-40	5.1	40-60	3.2
30-50	4.8	60-80	2.9
40-60	4.7	80-100	2.8
50-70	4.6	100-120	2.8
60-80	4.6	120-140	3.0
70-90	4.6	140-160	3.1
80-100	4.9		
90-110	5.8		

FUEL CONSUMPTION

Overall	18.9 mpg 14.9 litres/100 km
Touring*	22.5 mpg 12.5 litres/100 km
Govt tests	15.8 mpg (urban) 31.4 mpg (56 mph) 25.2 mpg (75 mph)
Fuel grade	97 octane 4 star rating
Tank capacity	47 litres 16.3 galls
Max range*	367 miles 590 km
Test distance	630 miles 1014 km

*Based on official fuel economy figures – 50 per cent of urban cycle, plus 25 per cent of each of 56/75 mph consumptions.

STEERING

Turning circle	9.4 m 30.7 ft
Lock to lock	3.1 turns

NOISE

	dBA
30 mph	71
50 mph	75
70 mph	83
Maximum†	94

†Peak noise level under full-throttle acceleration in 2nd

SPEEDOMETER (MPH)

True mph	30	40	50	60	70	80	90	100
Speedo	34	44	55	67	79	90	101	112

Distance recorder: 0.6 per cent fast

WEIGHT

	Kg	cwt
Unladen weight*	1325	26.1
Weight as tested	1530	30.1

*No fuel

Performance tests carried out by *Motor's* staff at the Motor Industry Research Association proving ground, Lindley, and Millbrook proving ground, near Ampthill.

GENERAL SPECIFICATION

ENGINE

Cylinders	V8
Capacity	3185 cc
Bore/stroke	83/73.6 mm
Max power	270 bhp (198.6 kW) at 7000 rpm (DIN)
Max torque	224 lb ft (304 Nm) at 5500 rpm (DIN)
Block	Aluminium alloy
Head	Aluminium alloy
Valve gear	Dohc per bank, 4 valves per cylinder
Compression	9.8:1
Fuel system	Bosch K-Jetronic fuel injection
Ignition	Marelli Microplex
Bearings	Five main

TRANSMISSION

Drive	To rear wheels
Type	Five-speed manual
Internal ratios and mph/1000 rpm	
Top	0.919:1/21.0
4th	1.244:1/15.5
3rd	1.693:1/11.4
2nd	2.353:1/8.2
1st	3.419:1/5.6
Rev	3.248:1
Final drive	3.70:1

SUSPENSION

Front	Independent by double wishbones, coil springs, anti-roll bar.
Rear	Independent by double wishbones, coil springs, anti-roll bar.

STEERING

Type	Rack and pinion
Assistance	None

BRAKES

Front	Vent discs 27.4 cm dia
Rear	Vent discs 24.9 cm dia
Servo	Yes
Circuit	Split front/rear
Rear valve	No

WHEELS/TYRES

Type	Light alloy, 7 × 16 in/ 8 × 16 in dia
Tyres	205/55 VR 16, 225/50 VR 16
Pressures F/R (normal)	33/33 psi 2.3/2.3 bar

ELECTRICAL

Battery	12 v, 66 Ah
Alternator	85 Amp
Fuses	23
Headlights	
type	Halogen
dip	110 W total
main	120 W total

GUARANTEE

Duration	12 months unlimited mileage.
Rust warranty	Coupon plus yearly checks

MAINTENANCE

Major service	7000 miles

Make: *Ferrari* ***Model:*** *328 GTB* ***Country of Origin:*** *Italy*
Maker: *Ferrari Esercizio Fabbriche Automobilie Corse SpA, Modena.*
UK Concessionaire: *Maranello Sales Limited, Egham By-Pass (A30), Egham, Surrey TW20*
Tel: *Egham 36431* ***Total Price:*** *£34,750* ***Options:*** *Air conditioning (£1183.55)*

THE RIVALS

Other rivals include the De Tomaso Pantera GT5 5.8 (£34,247), Mercedes 500 SEC (£40,400) and Porsche 928 S2 (£38,519)

FERRARI 328 GTB

£34,750

56% | 44%

Length 4·25m (167·3") | Width 1·73m (68·1") | Front track 1·48m (58·3")
Wheelbase 92·5m (2·35") | Height 1·13m (44·4") | Rear track 1·46m (57·5")

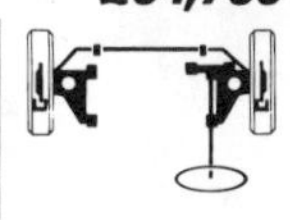

Capacity, cc	3185
Power bhp/rpm	270/7000
Torque lb ft/rpm	224/5500
Max speed, mph	158.5
0-60 mph, sec	5.5
30-50 mph in 4th, sec	4.8
mph/1000 rpm	21.0
Overall mpg	18.9
Touring mpg	22.5
Weight kg	1325
Drag coefficient Cd	N/A
Boot capacity m³	0.16

328 supersedes 308 QV and is more desirable still. With a top speed of around 160 mph and scintillating mid-range acceleration, outright performance now eclipses 911 Carrera's, though engine is very loud and sounds disappointingly Fiat-like at low revs. Handling is impeccable and ride supple, but power can be used to out-gun grip and chassis won't easily forgive wild slides. Still a beautiful car, minor cabin changes have improved habitability.

AUDI QUATTRO

£24,204

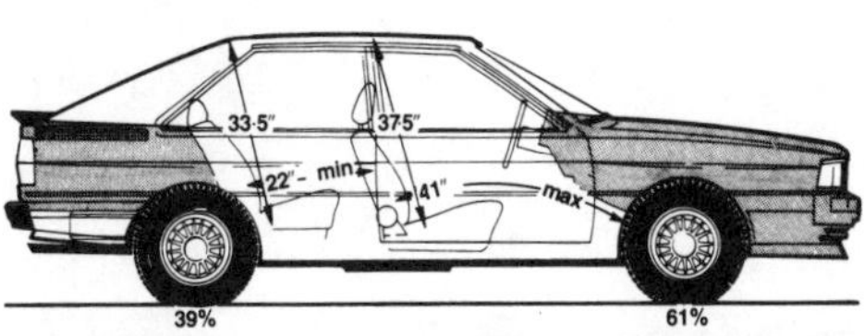

39% | 61%

Length 4·40m (173·3") | Width 1·72m (67·8") | Front track 1·45m (57·3")
Wheelbase 2·52m (99·3") | Height 1·35m (53") | Rear track 1·42m (56")

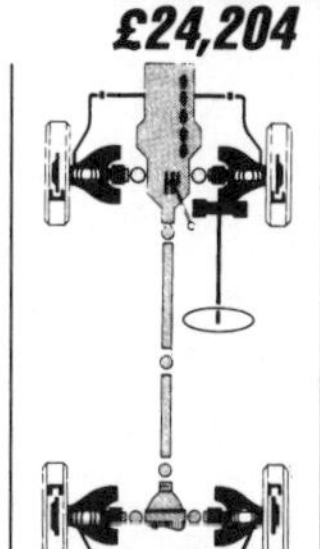

Capacity, cc	2144
Power bhp/rpm	200/5500
Torque lb ft/rpm	210/3500
Max speed, mph	138e
0-60 mph, sec	6.5
30-50 mph in 4th, sec	8.2
mph/1000 rpm	23.6
Overall mpg	19.9
Touring mpg	24.7
Weight kg	1260
Drag coefficient Cd	0.43
Boot capacity m³	0.21

*Figures for old model

A more civilised machine these days than its epoch-making predecessor, the 4wd Quattro is more than ever a milestone in car design. It combines phenomenal roadholding and traction with performance, refinement, economy, comfort and accommodation in a way that has no equal, against which its weaknesses – poor ratios (still) and slow shift, unprogressive heating, sparse instruments – are minor failings. Now available with anti-lock braking.

BMW M635 CSi

£37,750

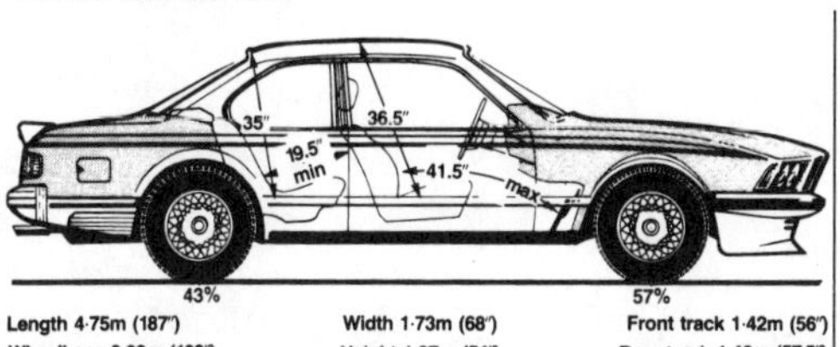

43% | 57%

Length 4·75m (187") | Width 1·73m (68") | Front track 1·42m (56")
Wheelbase 2·62m (103") | Height 1·37m (54") | Rear track 1·46m (57·5")

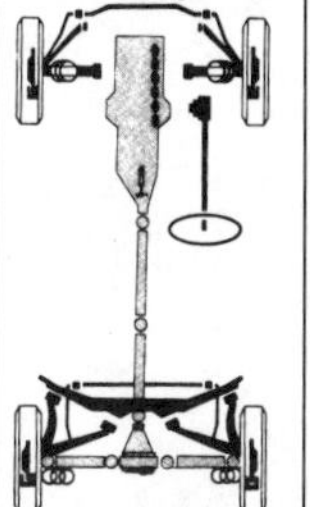

Capacity, cc	3453
Power bhp/rpm	286/6500
Torque lb ft/rpm	251/4500
Max speed, mph	149.7
0-60 mph, sec	6.3
30-50 mph in 4th, sec	6.7
mph/1000 rpm	23.8
Overall mpg	16.9
Touring mpg	24.0
Weight kg	1506
Drag coefficient Cd	N/A
Boot capacity m³	0.35

Fastest BMW available in the UK takes the decade-old 6-Series firmly into supercar territory and more than holds its own. Fabulous 24-valve six delivers lusty performance with good refinement but is very thirsty. Handling is both enjoyable and forgiving, ride firm but well-controlled. Interior is too ordinary for £37,000 but build and finish are first class. Generally well equipped but air conditioning is extra.

LAMBORGHINI JALPA 350

£36,577

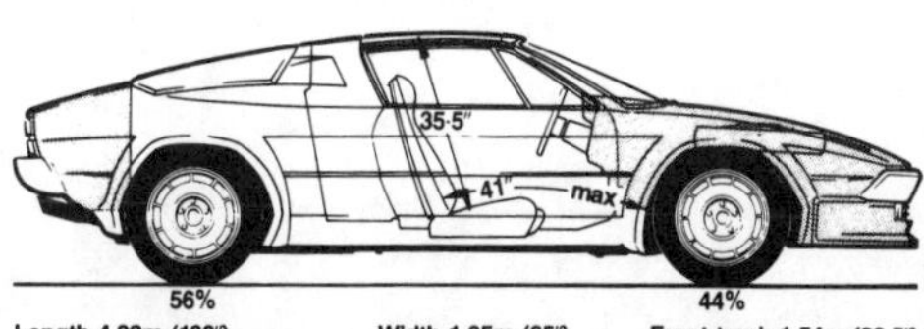

56% | 44%

Length 4·22m (166") | Width 1·65m (65") | Front track 1·54m (60·5")
Wheelbase 2·45m (96·5") | Height 1·12m (44") | Rear track 1·48m (58·5")

Capacity, cc	3485
Power bhp/rpm	250/7000
Torque lb ft/rpm	235/3250
Max speed, mph	147.6
0-60 mph, sec	5.8
30-50 mph in 4th, sec	4.3
mph/1000 rpm	20.4
Overall mpg	15.6
Touring mpg	–
Weight kg	1351
Drag coefficient Cd	–
Boot capacity m³	–

"Baby" of the two-car Lamborghini range, the targa-top Jalpa's ancestry runs back to the mid-engined Uracco of the early '70s. Magnificently vocal quad-cam V8 delivers fine performance with massive mid-range punch, though economy is mediocre by today's standards. Very safe and ultimately forgiving handling married to reasonable ride. Fabulous brakes. He-man gearchange and poor visibility not so appealing, but practicality and finish better than of yore.

LOTUS ESPRIT TURBO

£23,440

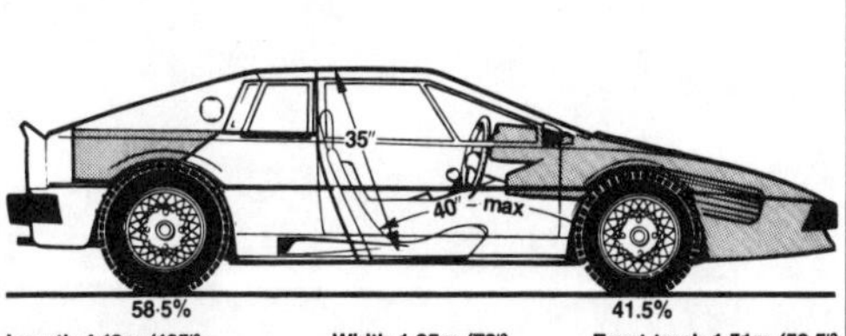

58·5% | 41.5%

Length 4·19m (165") | Width 1·85m (73") | Front track 1·51m (59·5")
Wheelbase 2·44m (96") | Height 1·11m (43·8") | Rear track 1·51m (59·5")

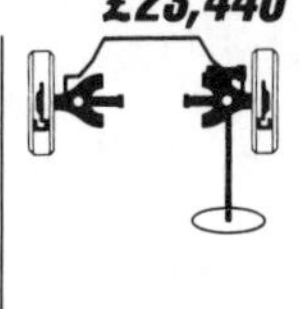

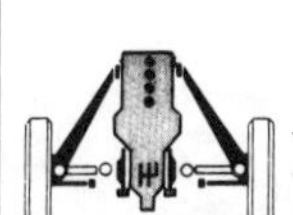

Capacity, cc	2172
Power bhp/rpm	210/6500
Torque lb ft/rpm	200/4500
Max speed, mph	152*
0-60 mph, sec	5.6
30-50 mph in 4th, sec	6.2
mph/1000 rpm	22.7
Overall mpg	18.5
Touring mpg	27.9
Weight kg	1148
Drag coefficient Cd	0.33
Boot capacity m³	0.10

*estimated

Turbo power promotes Lotus towards the top of the first division in the supercar league. Stunning acceleration from smooth and vice-less turbo "four" allied to perfect ratios, strong roadholding, superb handling and tireless braking, all adding up to a driver's car par excellence, with respectable economy and a comfortable ride as icing on the cake. But some shortcomings include the poor heating, visibility, pedal layout and lack of space for tall drivers.

PORSCHE 911 CARRERA SPORT

£36,676

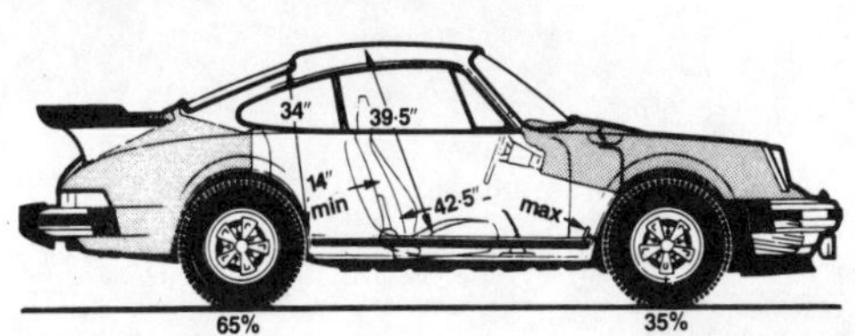

65% | 35%

Length 4·29m (169") | Width 1·77m (69·8") | Front track 1·50m (59")
Wheelbase 2·27m (89·5") | Height 1·32m (52") | Rear track 1·43m (56.3")

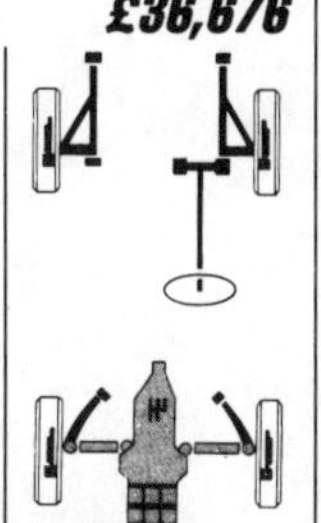

Capacity, cc	3164
Power bhp/rpm	231/5900
Torque lb ft/rpm	209/4800
Max speed, mph	151.1
0-60 mph, sec	5.3
30-50 mph in 4th, sec	5.6
mph/1000 rpm	24.3
Overall mpg	21.1
Touring mpg	28.6
Weight kg	1166
Drag coefficient Cd	0.38
Boot capacity m³	0.28

Little changed for the past couple of years but still at the top of the junior supercar acceleration league table – the 911 Carrera is also remarkably economical for its stunning performance. Still a great driving machine, with rewarding (though tricky *in extremis*) handling, potent brakes, superb ratios, good driving position and turbine-smooth engine. Gear change could be better, though, and remaining flaws include hard ride, and poor heating/ventilation.

FERRARI 328 GTS

Refining an exotic legend

PHOTOS BY JOHN LAMM

THERE ARE MANY classes of cars, from economy sedans to sports and GT cars, and atop the heap for most enthusiasts are the exotic cars from Ferrari, Lamborghini and others. Only a tiny percentage of us will ever own an exotic, but they represent the dream car for most of us, and it's enough to see one up close and hear the lovely sounds that most of these cars broadcast as part of their mystique. It may seem, therefore, something of a contradiction to talk of the Ferrari 328 GTS as an honest everyday driving car, given the exotics' reputation for being temperamental, but this updated version of the popular 308 is among the first cars from Ferrari with performance designed for the American market.

The feature that makes the 328 particularly interesting—and that puts it into the everyday-driving spectrum—is the engine. During the past 40 years, Ferrari has built 165 types of engines, and for the past 10 years 75 percent of those have been V-8s. The previous 308 was powered by a 2927-cc V-8 with four valves per cylinder and had a bore and stroke of 81.0 x 71.0 mm. With the advent of the 328, the displacement has grown to 3185 cc, with increases to both bore and stroke: 83.0 x 73.6 mm. There is also a new piston design with a pronounced squish effect for improved combustion, a higher compression ratio (9.2:1 versus 8.8), higher-profile camshafts and larger oil radiators. Horsepower and torque are up 13.0 and 13.3 percent, respectively.

From the instant you fire up the engine and start driving the car, these changes are apparent. The 328 powerplant rewards with its enormous flexibility. This is an engine that will motor sedately around town, making soothing rumbling noises, or will wind itself up to the redline with that exciting Ferrari snarl that has echoed around every major race circuit in the world.

With 260 bhp at 7000 rpm and 213 lb-ft of torque at 5500, this is a car that will charge off from a standing start and get itself to 60 mph in a mere 6.0 seconds. And through the quarter mile, it needs only 14.5 sec to flash through the lights at an impressive 96.0 mph. This level of performance is not available in most cars, but then most cars don't cost $63,370 either.

Yet this is also an engine that is not the least upset if the driver trundles around town in 5th gear at engine speeds as low as 1000 rpm. When asked to perform, the 328 pulls from a crawl smoothly and utterly without fuss. One driver commented that this car and

AT A GLANCE	Ferrari 328 GTS	Lotus Turbo Esprit	Porsche 911 Turbo
Price, base/ as tested	$63,370 $63,370	$54,942 $54,942	$48,000 $49,142
Curb weight, lb	3170	2710	3060
Engine/drive	V-8/rwd	inline-4/rwd	flat-6/rwd
Transmission	5-sp M	5-sp M	4-sp M
0–60 mph, sec	6.0	6.6	5.0
Standing ¼ mi, sec @ mph	14.5 @ 96.0	15.3 @ 90.0	13.4 @ 103.0
Stopping distance from 60 mph, ft	154	154	143
Lateral acceleration, g	0.83	0.81	0.84
Slalom speed, mph	60.3	62.2	na
Fuel economy, mpg	16.5	17.5	17.0

	Pro	Con
328 GTS:	a Ferrari is a Ferrari, right?, good acceleration, updated looks, very good finish	high price, oversteer in abrupt transitions, little head room, awkward steering wheel
Turbo Esprit: tested 12-83	an Englishman in a world of Italian exotics, improved fit and finish, good torque for a turbo	a little floaty at speed, sloppy shifter, should any 4-cylinder cost this much?
911 Turbo: tested 1-86	fearsomely fast if you keep the pedal down, ageless design, high quality finish	substantial turbo lag, sloppy shifter, delicate handling at the limit (still)

the Testarossa are the first Ferraris designed with U.S. driving conditions in mind from the outset.

However, getting the most out of this lovely V-8 engine demands patience, precision and persistence on the part of the driver because of transmission "problems." The gearshift lever is a thin steel rod with a comfortable knob on top, but the right-out-there-for-anyone-to-see shift gate will slow down even the most accomplished driver until he's had lots of practice. And, should the temperature of the gearbox oil be less than spa-warm, gear changes take on the character of a paddle moving through very thick glue. On moderately cold mornings here in southern California—and we're only talking 40 degrees Fahrenheit or so—2nd gear becomes virtually unobtainable for about 10-minutes running unless you're practiced in double-clutch upshifting as well as downshifting. Once the lubricant is up to operating temperature, however, the fun returns: You're tickling that transmission up and down through the gears while negotiating the Targa Florio or Le Mans, not driving the parkway to work. And all the while that terrific and flexible powerplant is responding instantly to throttle inputs.

As for the steering, a racing heritage goes hand in leather glove with a rack-and-pinion system, and the 328's is quite precise. Low-speed effort is on the heavy side, but once you have the car moving beyond a walking pace, the effort is just fine. The response is a bit slower than we would like and the car feels like it needs more caster built in because the steering doesn't center itself easily. It's also not quick in recovering after a turn. Another aspect we found annoying was the steering wheel angle: It's canted too far forward at the top, making it tough for the driver to react quickly when necessary, such as in our 700-ft slalom test.

The Ferrari performed this exercise adequately, at a speed of 60.3 mph, but you do have to work at it and the wheel angle is not your ally. Nor is the rear end, which can be a real handful in quick transitions. The 328 circles the 200-ft diameter skidpad with a lateral acceleration of 0.83g, respectable, but certainly not outstanding these days.

On any sort of road, smooth or bumpy, however, this is a car that will lope through high-speed bends effortlessly, and tackle the tight turns with basic understeer until you near the limit. Then the rear-weight bias takes over and the 328, like many mid-engine cars, becomes an oversteerer. Some of this reaction, as well as the rather average skidpad performance, is attributable to the relatively narrow tires, 225/50VR-16 at the rear, 205/55s in the front.

Certainly most drivers rarely approach these limits on the road, but will instead delight in the supple ride characteristics and the precise cornering ability. This is a car that is great fun to drive under any condition, and is particularly suited to those winding country lanes we all treasure.

The 4-wheel vented disc brakes inspire confidence up to the point of an all-out, panic stop. At the test track, though, we discovered our test car had locking tendencies for both right wheels, causing the driver to have to work to keep the car in a straight line. For that reason, our stopping distances from 60 and 80 mph are not exceptional—154 and 265 ft, respectively. An evidently fixable situation aside,

The 328's restyled interior has electric window switches behind door pull (above) and separate ventilation controls for driver and passenger on center console (right).

we feel a car of this caliber deserves ABS.

As with other exotics, the Ferrari 328 is built low to the ground, so getting in and out of the driver's seat is not like jumping into the family sedan. The two seats are covered with leather and look beautiful. They are newly designed for the 328, with more padding and lateral support, but they don't hold you firmly in place during hard cornering. Taller drivers will find head room skimpy (Tom Selleck of TV's *Magnum, P.I.* must never put the roof panel in place) and the reach to the steering wheel rather a long one.

The instrument panel is clearly visible, except for some warning lights blanked out by the steering wheel, and the gauges are big enough to read at a glance. Most of the controls and switches are easy to use, but the heat/vent/air conditioning system is new for the 328 and requires some explanation or careful reading of the owner's manual. There are separate controls for driver and passenger for heat and vent, and they are activated via rocker switches on the console. Each person can set direction of airflow, temperature and fan speed. The air conditioning, however, has only a single set of controls. Large round outlets on the top of the dash and two small rectangular ones on the facia can direct air to both occupants and will perform adequately except on the very hottest days.

There is no glovebox to speak of, but there are small compartments in each door and between the seats. The luggage compartment is also limited in size and is accessible by opening the rear hatch, which also puts the gorgeous engine on display for bellhops. The front compartment is pretty well filled with the temporary spare tire and other mechanical items.

Ferrari has apparently gone to great trouble to redesign the door latches inside and out, but we found the inside ones awkward and rather devilishly well hidden from the first-time driver.

But let's put aside the minor complaints and refocus on the car as a whole. It looks terrific, thanks to the redesigned front end with integrated bumper (body color now) and the larger air dam. At the rear, too, the bumper blends nicely into the body so there is no longer that abrupt transition from black bumper to red (or whatever other color you choose) paint. The overall effect is considerably more graceful than the 308's lines were, and that car has been generally admired.

The 328 is a fine example of evolutionary improvement and this latest version of Ferrari art is a wonderful driving machine. To those who are fortunate enough to afford rolling art, the engine's flexibility and performance are worth the price of admission. And then there's the plus of driving a Ferrari—a legend when enthusiasts dream about exotic cars.

There's little room for anything extra in front compartment. The 328's V-8 engine is beautiful to look at and boasts significant bhp and torque increases.

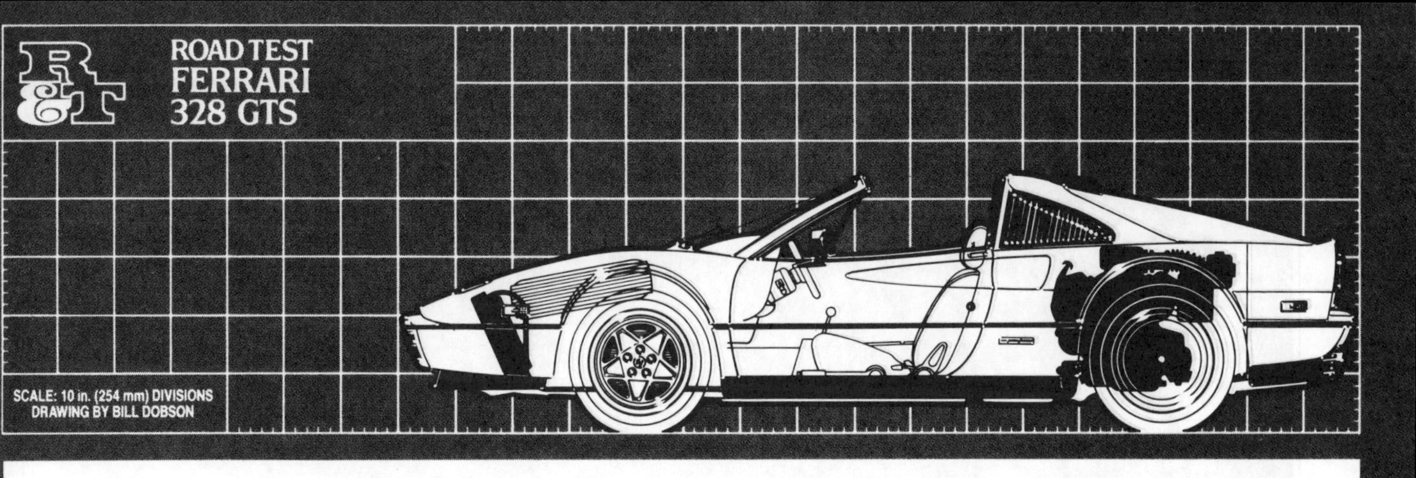

PRICE

	328 GTS 1986	308 GTSi 1983
List price, all POE	$63,370	$53,745
Price as tested	$63,370	$53,745

Price for 328 GTS as tested includes std equip. (air cond, elect. window lifts, elect. mirrors)

IMPORTER

Ferrari North America, PO Box 413, Montvale, N.J. 07645

GENERAL

	328 GTS	308 GTSi
Curb weight, lb	3170	3250
Test weight	3340	3440
Weight dist (with driver), f/r, %	44/56	42/58
Wheelbase, in.	92.5	92.1
Track, f/r	58.0/57.8	57.8/57.8
Length	168.7	174.2
Width	68.1	67.7
Height	44.4	44.1
Ground clearance	4.0	3.0
Overhang, f/r	42.0/34.2	45.0/37.1
Trunk space, cu ft	5.3[1]	
Fuel capacity, U.S. gal.	18.5	

[1]Single entries indicate specifications are identical.

ENGINE

	328 GTS	308 GTSi
Type	dohc 4-valve V-8	
Bore x stroke, mm	83.0 x 73.6	81.0 x 71.0
Displacement, cc	3185	2927
Compression ratio	9.2:1	8.8:1
Bhp @ rpm, SAE net	260 @ 7000	230 @ 6800
Torque @ rpm, lb-ft	213 @ 5500	188 @ 5500
Fuel injection	Bosch K-Jetronic	
Fuel requirement	unleaded, 91-pump oct	

Exhaust-emission control equipment: 3-way catalyst with oxygen sensor, air injection

DRIVETRAIN

	328 GTS		308 GTSi	
Transmission	5-sp manual			
Gear ratios: 5th	(0.92)	3.51:1	(0.92)	3.74:1
4th	(1.24)	4.74:1	(1.24)	5.03:1
3rd	(1.69)	6.46:1	(1.69)	6.86:1
2nd	(2.35)	8.98:1	(2.35)	9.54:1
1st	(3.39)	12.95:1	(3.42)	13.89:1
Final drive ratio		3.82:1		4.06:1

CALCULATED DATA

	328 GTS	308 GTSi
Lb/bhp (test weight)	12.8	14.9
Bhp/liter	81.6	78.6
Engine revs/mi (60 mph in top)	3100	3300
R&T steering index	1.30	
Brake swept area, sq in./ton	261	

CHASSIS & BODY

	328 GTS	308 GTSi
Layout	transverse mid-engine/rear drive	
Body/frame	steel/tubular steel	
Brake system, f/r	10.7-in. vented discs/ 10.9-in. vented discs; vacuum assist	
Wheels	cast alloy, 16 x 7J	cast alloy, 390 x 190
Tires	Goodyear NCT, 205/55VR-16 f 255/50VR-16 r	Michelin TRX, 220/55VR-390
Steering type	rack & pinion	
Turns, lock-to-lock	3.3	
Turning, circle, ft	39.4	

Front suspension: unequal-length A-arms, coil springs, tube shocks, anti-roll bar

Rear suspension: unequal-length A-arms, coil springs, tube shocks, anti-roll bar

INSTRUMENTATION

Instruments: 180-mph speedometer, 10,000-rpm tach, oil press., coolant temp, oil temp, fuel level

ACCOMMODATION

	328 GTS	308 GTSi
Seating capacity, persons	2	
Head room, in.	34.5	35.5
Seat width	2 x 18.0	2 x 19.0
Seatback adjustment, deg	30	

ROAD TEST RESULTS

ACCELERATION

	328 GTS	308 GTSi
Time to distance, sec:		
0–100 ft	3.0	3.0
0–500 ft	7.9	8.2
0–1320 ft (¼ mi)	14.5	15.2
Speed at end of ¼ mi, mph	96.0	91.5
Time to speed, sec:		
0–30 mph	2.0	2.3
0–60 mph	6.0	6.8
0–80 mph	9.9	11.3
0–100 mph	15.9	18.1

SPEEDS IN GEARS

	328 GTS	308 GTSi
Maximum engine rpm	7700	7700
5th gear, mph	149	142
4th	112	110
3rd	86	84
2nd	61	60
1st	41	41

FUEL ECONOMY

	328 GTS	308 GTSi
Normal driving, mpg	16.5	16.0

HANDLING

	328 GTS	308 GTSi
Lateral accel, 100-ft radius, g	0.83	0.81
Speed thru 700-ft slalom, mph	60.3	60.9

BRAKES

	328 GTS	308 GTSi
Minimum stopping distances, ft:		
From 60 mph	154	153
From 80 mph	265	262
Control in stop	poor	very good
Pedal effort for 0.5g stop, lb	22	
Fade: percent increase in pedal effort to maintain 0.5g deceleration in 6 stops from 60 mph	nil	
Overall rating	fair	very good

INTERIOR NOISE

	328 GTS	308 GTSi
Idle in neutral, dBA	69	73
Max, 1st gear	88	83
Constant 30 mph	74	74
50 mph	76	77
70 mph	77	80
90 mph	80	na

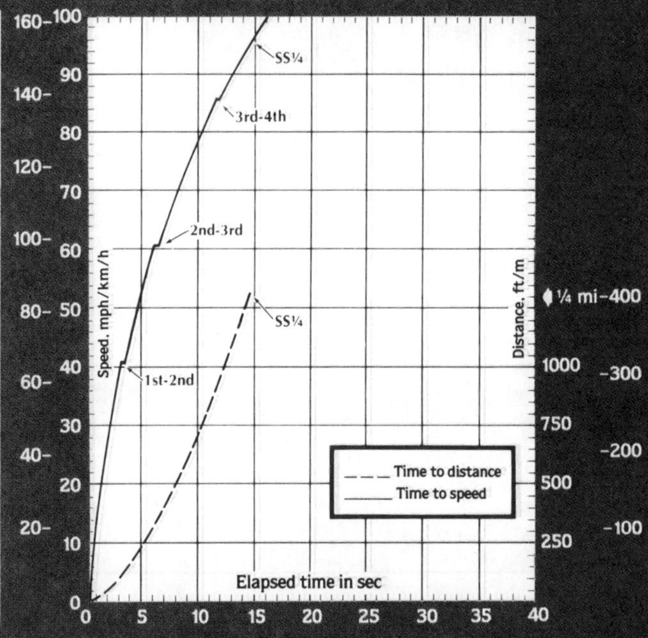

ON TOP OF THE WORLD

The 2+2 Mondial is the 'Ferrari for the family man.' Now with a more powerful engine, how does it shape up?

FERRARI MONDIAL 3.2 QV

FOR:

PERFORMANCE

APPEARANCE

AGAINST:

GEARING

ACCOMMODATION

There is no need to search for the name. The silver horse, rearing up on its hind legs, simply says it all: Ferrari. Park the car in a city street and watch just how many people pause to take a closer look at it. On the open road, and even if you are in no particular hurry, other drivers will make room for the car. And few drivers in GTi-badged hatchbacks will pit their cars against the 270bhp of thoroughbred Italian engineering.

The Mondial first appeared at the Geneva show six years ago, a 2+2 to complement the two-seater (then called the 308GTB). As with all the other current Ferraris, the Mondial has a tubular steel chassis and, like the 328, the body was designed by Pininfarina, although the cars are built at Ferrari's Modena factory.

The original 3-litre V8 engine had two valves per cylinder, but in 1982 the four-valve heads were fitted to make compliance with the US Federal emissions regulations easier. Last year, the capacity was raised to 3.2 litres and the power output increased to 270bhp at 7000rpm. The engine itself is a classic example of how to make a piece of engineering not only work efficiently and sound wonderful but also look the part. The red air intake plenum chamber, with its bright, machined fins, bears that evocative name — Ferrari.

Belt drives are used to the twin overhead camshafts, and the all-alloy block has wet cylinder liners. There is an oil cooler, with its own fan, located behind the air intake just ahead of the right-hand rear wheel. The water radiators, with twin electric fans, are at the front of the car, behind the three-slat grille.

PERFORMANCE

To anyone who is sitting behind the wheel of a Ferrari for the first time, the prospect of coping with all that potential power might seem a little daunting. But in practice the Mondial is an amazingly easy car to drive, thanks in no small way to the flexibility of that V8 engine and the curiously low gearing.

The Bosch K-Jetronic petrol injection gave first-time starting even when the engine was very hot after a long, high-speed run. Cold starting too, admittedly during some fairly warm weather, gave no problems. But until the gearbox had warmed up a little, it could be difficult to select second gear. The gearchange pattern follows the classic racing type

TECHNICAL FOCUS

When the 2+2 Mondial first appeared at the 1980 Geneva Show, it was a car with the United States market very much in mind. At that time 35 per cent of Ferrari production went to the States, and the Mondial was designed to meet all forthcoming safety regulations which had just been announced.

The initial engine size was 3-litre, the same as that in the 308GTB, but with Bosch K-Jetronic injection in place of the latter's Weber carburettors. Power output for that engine was 214bhp at 6500rpm. Two years later, the Mondial engine was given new cylinder heads, with four rather than two valves per cylinder — hence the qv for *quattrovalvole*. The power was then hiked to 240bhp. This increase was necessary to keep up with the amount of power being absorbed by the US Federal requirements on exhaust emissions.

The last change came at the 1985 Frankfurt Show. The capacity was increased to 3.2-litres, with a corresponding boost in power, up to 270bhp.

In addition to the 2+2 coupé version, there is also a cabriolet model. Both cars are designed by Pininfarina, but built at Ferrari's Modena factory.

◀ with first (needed only for starting from a standstill) on its own down to the left and opposite reverse. The pencil-thin, chromium-plated gear lever, with a large, black spherical knob, working in a six-slotted gate, is of a design which has hardly changed since the days of the legendary 250GTO.

It does not take long to realise just how low-geared the Mondial is. Top gear pulls a mere 20.9mph/1000rpm, which means that the car can be trickled around in traffic in this gear, and then rocket up close to a glorious-sounding 7000rpm and 144mph.

Getting the Mondial off the line needed a little care. The fat Michelin 225/55VR16 Michelin TRX rear tyres are reluctant to spin, and unless the clutch is treated carefully the engine revs can be too easily killed. But once the technique has been mastered, the Mondial rockets away. It took just 2.6secs to reach 30mph, with 60 coming up in 6.8secs, 100 in 16.5secs and 130mph in 32.8secs.

The only slightly awkward movement of the gearchange is the dog-leg one from first to second. The others are simply a matter of moving the lever as quickly as you can, with quite a lot of effort needed to overcome the very powerful synchromesh.

A look at the in-the-gears figures shows just what a broad power and torque spread this V8 engine has. The power just comes pouring in, without any of the unseemly, sudden rush which is a feature of so many high-performance turbocharged engines.

The rev counter, which reads to 10,000rpm, is red lined at 7600rpm, but on the test car the rev limiter cut in at 7400rpm. The rev counter also under-read by 7.5 per cent. With its low gearing, it did not take the Mondial any great time or distance to reach its top speed of 144mph, equivalent to 6900rpm, or within a whisker of the top of the power curve.

ECONOMY

For anyone considering a car in the performance catergory of the Mondial, the cost of the fuel on which to run it will probably not appear very high on the list of priorites. Anything over the 20mpg mark is outstanding, but considering just how heavy the Mondial is, the 16.8mpg we achieved is reasonably good. Much of the test mileage was covered at well over 100mph, so under this country's rather stricter overall speed limit, the average owner could expect to get consumption figures rather better than that.

The 19.8-gallon tank has its filler under an electrically controlled flap on the right-hand rear wing. If the electrics fail, a toggle reached from the engine bay can be used in an emergency. Even when the car is being driven hard, it can be expected to cover around 300 miles between refills, allowing for a couple of gallons in reserve. Ferrari's years of racing experience mean that the Mondial's tank can be filled very easily, without any blow back or bubbling. Over the relatively short test distance, the V8 engine used no measurable amount of oil.

REFINEMENT

It would be a soulless character who did not revel in the wonderful noise which the Mondial's V8 engine creates when hard at work. Yet even when the rev counter needle is well past the 7000rpm mark, this wonderful piece of Italian engineering never sounds at all stressed. It was built to do just that sort of job.

Yet around town, the car is as docile as one could wish, the engine burbling softly. Its tractability also means that there is no need to go screeming up through the rev range when the traffic lights turn green. Wind noise is very well controlled, and even with the sunroof open, it is possible to carry on a conversation, albeit with raised voices.

There is some tyre noise, especially when the car is driven over coarse, concrete surfaces. But for a car of such high performance, the noise levels are generally very well controlled.

ROAD BEHAVIOUR

In a world where practically every other car seems to be nose heavy, the Mondial comes as a refreshing surprise, with 55.4 per cent of its weight being carried over the rear wheels. The suspension follows classic lines — double wishbones front and rear, with coil springs and telescopic dampers. There are also anti-roll bars and rear.

At low speeds, the rack and pinion steering can be quite heavy, especially when parking the Mondial in tight spots. But as speed builds, the heaviness disappears, leaving the driver knowing exactly what the front wheels are doing. Three and a half turns are needed to get from lock to lock, and the turning circle is a rather stately 41½ft. It is when the car is being driven on ordinary roads that the amount of bump steer can be felt, twitching the car slightly off line as it passes over manhole covers and other irregularities. This never amounts to anything serious, and the quickness of the steering makes this behaviour easily countered.

It is to the credit of Ferrari's chassis engineers and the aerodynamicists at Pininfarina that the Mondial feels absolutely rock steady at 140mph, with no need for any obvious air dams or spoilers.

The handling is virtually neutral, with perhaps a hint of understeer. With such a flexible engine, it *is* possible to move the rear wheels off line in a corner, but never suddenly. Despite its size, this is a very easy car to drive quickly. Ride quality is remarkably good, for one might expect over-stiff springing and rock-hard damping.

If there is criticism to be aimed at the braking system, it is that in town it can feel a little over sensitive, almost over-servoed. One soon gets used to being slightly feather-footed to prevent the car from being brought to an unseemly and snatchy halt. It takes only 20lb pedal pressure to achieve 0.36g — that is the sort of retardation on which passengers start to comment. And with just 30lb more pedal pressure, the car was brought up at 1.0g, with the fronts just starting to lock. With the majority of the car's weight over the fat-section rear wheels, it is not surprising that the handbrake, located on the outside of the driving seat, recorded an impressive 0.41g.

As one might expect from a car in this class, the brakes coped with our accelerated fade tests extraordinarily well, hauling the speed down at 0.5g from 95mph 10 times in succession without any problems, apart

Even the *relatively 'unsporting' Mondial 2+2 turns heads*

Cooling air *for the engine is ducted through inlets*

from a slight smell at the sixth stop and a hint of vibration on the ninth.

AT THE WHEEL

There is certainly nothing spartan or basic about the interior of the Mondial. Cream handstitched leather is used on the seats, facia and door panels. The slim-rimmed steering wheel has three spokes, with the familiar prancing horse motif appearing on the boss.

The instrument panel is mounted quite high so that minimum vertical eye movement is needed to look from road to instruments and back again. The large speedometer — reading 6mph fast at 100mph — is flanked by an impressive rev counter both with easy-to-read figures. Between these and the four smaller dials are the total and trip distance recorders, with a less than one per cent error. The smaller dials are for oil pressure and temperature, coolant temperature and fuel tank contents. The column levers follow a normal pattern, with a short one on the left for indicators. Behind this is the longer one for driving lamps. You cannot, incidentally, flash the Mondial's headlamps without first putting them up. The pedals are offset toward the middle of the car.

On our left-hand-drive test car, the gear lever, with its clearly marked gate, was to the left of the central console. On right-hand drive cars, it is moved to the right. Alongside the gear lever is a row of warning lamps covering such items as screen washer level, coolant level and stop lamp bulbs. Behind these there are the rocker switches for windows, and the locks for front and rear boot lids, engine cover, fuel filler flap and glove locker. This last seems a little unnecessary, as the amount of room provided behind the flap is minimal.

During the test period, we were driving in temperatures of over 27 deg C. But the Ferrari's air conditioning, which is a standard fitting, was barely able to keep cabin temperatures comfortable, even on the coldest setting. This was the one really disappointing feature of the car which is well appointed in so many ways.

CONVENIENCE

The Mondial is close to 6ft wide with the doors closed, so you need a fairly generous amount of room around the car when they are open. In practice, the angle at which they open is a little restricted and this, combined with the rather high sills, makes entry and exit a little awkward. The front seats are a little on the hard side, and lack much lateral support, a shortcoming noticed particularly when the car was being driven fast on twisting roads.

Rear seat leg and headroom is distinctly on the marginal side — but the children would no doubt appreciate arriving at school in a Mondial. The armrest between the rear seats contains a first-aid kit and conceals the emergency release for the engine cover.

Where the Mondial really does fall down is in the amount of room — or lack of it — for carrying odds and ends inside. The glove locker is practically useless and there are only small elasticated pockets on the doors for the handbook and a small map. The only place to stow cigarettes or sweets is in the small, slippery tray positioned alongside the gear lever.

The front 'boot' is not intended to carry much luggage. The brake servo, air conditioning accumulator and spare wheel take up much of the room. The spare is a 220/55VR390 Michelin TRX front wheel, which can also be used over a limited range on the rear. The main boot is located behind the engine bay, and holds a reasonable amount of luggage for two people.

Access around the engine is tight, but apart from some routine checking of oil and coolant levels, this is not the sort of car on which to indulge in too much DIY work.

It is not until you see someone sitting in the Mondial that you realise just how far forward the seats are located. This is turn means that visibility over that relatively long nose is quite good, making the car easy to place in crowded conditions. But care has to be taken when leaving angled side turnings, for there is a major blind spot created by the thick rear quarter pillars

To keep *eye movement to a minimum, instruments are mounted high*

Tremendous traction *is provided by Michelin 225/55VR16 tyres*

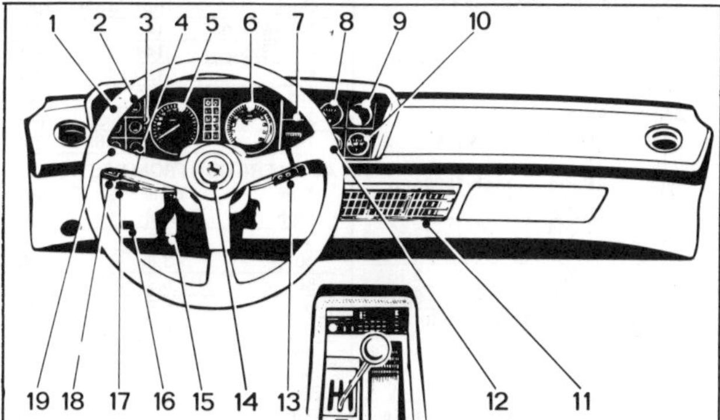

1 Heated rear window, **2** Front bonnet release, **3** Engine cover release, **4** Rear boot cover release, **5** Speedometer, **6** Rev counter, **7** Total and trip distance recorders, **9** Coolant temperature, **10** Oil temperature, **11** Centre air conditioning vent, **12** Oil pressure, **13** Screen wipers, **14** Horn push, **15** Column adjustment, **16** Air conditioning thermostat, **17** Direction indicators, **18** Driving lamps, **19** Front and rear fog lamps

ROAD TEST

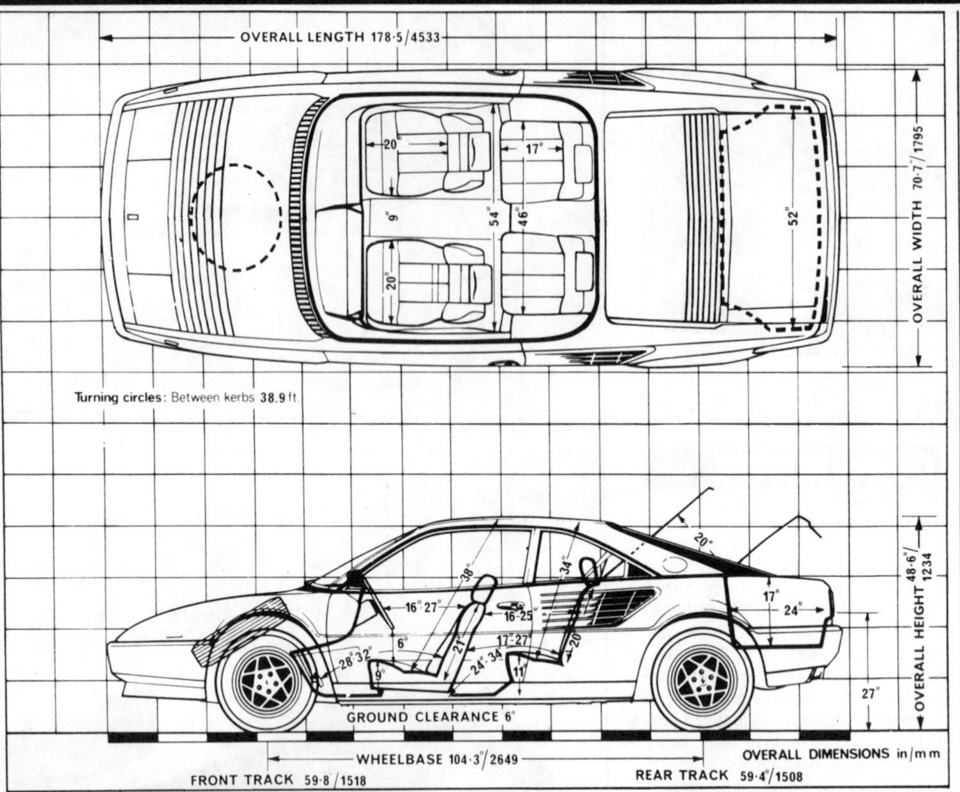

MODEL

FERRARI MONDIAL 3.2QV

PRODUCED BY:
Ferrari Esercizio Fabbrichee Corse SpA, Viale Trento Trieste 31, 41100 Modena, Italy

SOLD IN THE UK BY:
Maranello Concessionaires Ltd., Crabtree Road, Thorpe Industrial Estate, Egham, Surrey TW20 8RJ

SPECIFICATION

ENGINE
Transverse mid, rear-wheel drive. Head/block al. alloy/al. alloy. 8 cylinders in 90 deg V, wet liners, 5 main bearings. Water cooled, electric fan.

Bore 83mm (3.27in), **stroke** 73.6mm (2.90in), **capacity** 3186cc (194.5 cu in).

Valve gear 2 ohc, 4 valves per cylinder, toothed belt camshaft drive. **Compression ratio** 9.8 to 1. Marelli Microplex ignition, Bosch K-Jetronic carburettor.

Max power 270bhp (PS-DIN) (199kW ISO) at 7000rpm. **Max torque** 224lb ft at 5500rpm.

TRANSMISSION
5-speed manual. Single dry plate clutch.

Gear	Ratio	mph/1000rpm
Top	0.919	20.90
4th	1.244	15.44
3rd	1.693	11.35
2nd	2.353	8.16
1st	3.419	5.62

Final drive: hypoid bevel, ratio 3.823.

SUSPENSION
Front, independent, wishbones, coil springs, telescopic dampers, anti-roll bar.

Rear, independent, wishbones, coil springs, telescopic dampers, anti-roll bar.

STEERING
Rack and pinion. Steering wheel diameter 14.8in, 3.5 turns lock to lock.

BRAKES
Dual circuits, split diagonally. **Front** 11.1in (282mm) dia ventilated discs. **Rear** 11in (280mm) dia ventilated discs. Vacuum servo. Handbrake, side lever acting on rear discs.

WHEELS
Al. alloy, 7in rims front, 8in rear. Radial ply tyres (Michelin TRX on test car), size 220/55VR390 F, 240/55VR390 R, pressures F34 R35 psi (normal driving).

EQUIPMENT
Battery 12V, 66Ah. Alternator 85A. Headlamps 110/220W. Reversing lamp standard. 23 electric fuses. 2-speed, plus intermittent wipe screen wipers. Electric screen washer. Water valve interior heater; air conditioning standard. Leather seats, headlining.

PERFORMANCE

MAXIMUM SPEEDS

Gear	mph	km/h	rpm
Top (Mean)	143	230	6840
(Best)	144	232	6900
4th	120	193	7800
3rd	89	143	7800
2nd	64	97	7800
1st	43	69	7800

ACCELERATION FROM REST

True mph	Time (sec)	Speedo mph
30	2.6	32
40	3.8	42
50	5.3	52
60	6.8	63
70	8.9	75
80	10.8	85
90	13.9	96
100	16.5	106
110	20.1	118
120	25.5	129
130	32.8	140

Standing ¼-mile: 14.9sec, 95mph
Standing km: 27.4sec, 121mph

IN EACH GEAR

mph	Top	4th	3rd	2nd
10-30	—	—	4.3	3.2
20-40	8.5	6.1	4.0	2.9
30-50	7.8	5.6	3.8	2.7
40-60	7.6	5.5	3.6	2.7
50-70	8.1	5.5	3.7	—
60-80	8.5	5.5	3.9	—
70-90	9.0	5.6	—	—
80-100	9.9	5.9	—	—
90-110	10.9	6.6	—	—
100-120	11.8	8.2	—	—
110-130	13.1	—	—	—

CONSUMPTION

FUEL
Overall mpg: 16.8 (16.8 litres/100km) 3.70mpl

Autocar constant speed fuel consumption measuring equipment incompatible with fuel injection

Autocar formula: Driving and conditions	Hard 15.1mpg Average 18.5mpg Gentle 21.8mpg

Grade of fuel: Premium, 4-star (97 RM)
Fuel tank: 17.6 Imp galls (80 litres)
Mileage recorder: 1.0 per cent short
Oil: (SAE 10W/40) negligible

BRAKING
Fade (from 95mph in neutral)
Pedal load for 0.5g stops in lb

	start/end		start/end
1	20-30	6	30-40
2	20-25	7	35-40
3	20-30	8	30-35
4	25-35	9	20-30
5	25-40	10	20-30

Response (from 30mph in neutral)

Load	g	Distance
20lb	0.36	84ft
30lb	0.71	42ft
40lb	0.85	35.4ft
50lb	1.00	30.1ft
Handbrake	0.41	73ft

Max gradient: 1 in 4

CLUTCH Pedal 41lb; Travel 4in

WEIGHT
Kerb 29.2cwt/3265lb/1477kg (Distribution F/R, 44.6/55.4)
Test 32.6cwt/3647lb/165kg
Max payload 917lb/415kg

COSTS

Prices

Basic	£30,100.00
Special Car Tax	£2508.33
VAT	£4891.25
Total (in GB)	**£37,499.58**
Licence	£100.00
Delivery charge	£330.00
Number plates	£20.00
Total on the Road (excluding insurance)	**£37,949.58**
Insurance group	OA
EXTRAS (fitted to test car)	
Electric sunroof	£903.23
Total as tested on the road	**£38,852.81**

SERVICE & PARTS

Change	Interval 6250	12,500	18,750
Engine oil	Yes	Yes	Yes
Oil filter	Yes	Yes	Yes
Gearbox oil	No	Yes	No
Spark plugs	Yes	Yes	Yes
Air cleaner	No	Yes	No
Total cost	£244.49	£622.50	£244.49

(Assuming labour at £23.00 an hour inc VAT)

PARTS COST (inc VAT)

Brake pads (2 wheels) front	£36.23
Brake pads (2 wheels) rear	£33.81
Exhaust complete	£1003.95
Tyre—each (typical) front	£124.03
rear	£135.99
Windscreen	£345.00
Headlamp unit	£60.95
Front wing	£233.45
Rear bumper	£621.00

WARRANTY
12 months/unlimited mileage

EQUIPMENT

Ammeter/Voltmeter	N/A
Automatic	N/A
Cruise control	N/A
Limited slip differential	●
Power steering	N/A
Trip computer	N/A
Steering wheel rake adjustment	●
Steering wheel reach adjustment	●
Self-levelling suspension	N/A
Headrests front/rear	●/N/A
Heated seat	N/A
Height adjustment	N/A
Lumbar adjustment	N/A
Rear seat belts	●
Seat back recline	●
Seat cushion tilt	N/A
Door mirror remote control	●
Electric windows	●
Heated rear window	●
Interior adjustable headlamps	N/A
Tinted glass	●
Headlamp wash/wipe	N/A
Central locking	●
Cigar lighter	●
Clock	●
Fog lamps	●
Internal boot release	●
Locking fuel cap	●
Metallic paint	●
Radio/cassette	●
Aerial	●
Speakers	●

● Standard ○ Optional at extra cost N/A Not applicable † Part of option package DO Dealer option

TEST CONDITIONS

Wind: 3-5mph
Temperature: 82deg C (28deg F)
Barometer: 30.1in Hg (1019mbar)
Humidity: 90per cent
Surface: dry asphalt and concrete
Test distance: 679miles

Figures taken at 8360 miles by our own staff on the Continent.

THE OPPOSITION

BMW M635 CSi — £37,750

Not in the same performance class as the true supercars, but close to it and practically preferable, with the perfection of what BMW is best at — the six-cylinder engine — in its ultimate 24-valve development in a comfortable, elegant, easy to see-out-of coupé. Excellent steering and brakes combined with roadholding more predictable than usual for BMW

Tested	28 Apr 1984
ENGINE	3453cc
Max Power	286bhp at 6500rpm
Torque	251lb ft at 4500rpm
Gearing	23.87mph/1000rpm
WARRANTY	12/UL, 6 anti-rust
Insurance Group	9
Automatic	N/A
5-Speed	●
Radio/cassette	DO
Sunroof	●
WEIGHT	3322lb

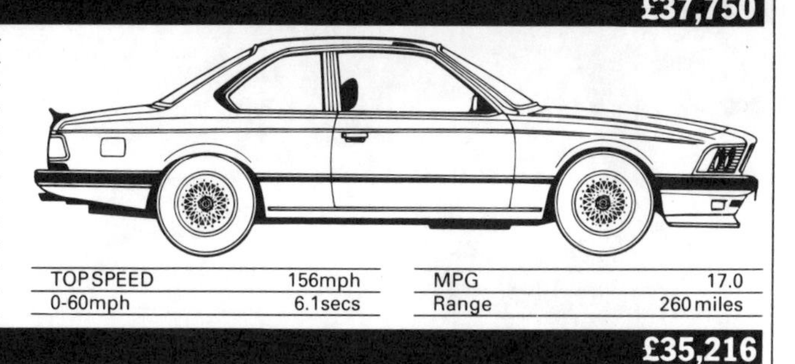

TOP SPEED	156mph	MPG	17.0
0-60mph	6.1secs	Range	260 miles

DE TOMASO PANTERA GT5 S — £35,216

Last of the American-engined Italian 'supercars', with design dating back to 1970. Mid-mounted V8 Ford engine has capacity of 5.8 litres and is available in three stages of power from 270 to 300bhp. Safe, predictable handling, but steering rather heavy at low speeds. Massive ventilated discs give massive stopping power. Seating limited to two only

Tested	N/A
ENGINE	5763cc
Max Power	270bhp at 5600rpm
Torque	325lb ft at 4000rpm
Gearing	26.4mph/1000rpm
WARRANTY	12/12,000, 5 anti-rust
Insurance Group	OA
Automatic	N/A
5-Speed	●
Radio/cassette	DO
Sunroof	N/A
WEIGHT	3138lb

*TOP SPEED	155mph	*MPG	14.1
*0-62mph	6.0secs	Range	290 miles

JAGUAR XJ-5 HE — £24,995

The XJ-S has always stood for effortless performance, the superbly-refined engine setting the car apart from any other on the road. For such a big beast, it is quite a nimble machine even round country lanes, thanks to responsive steering and viceless handling. Fuel consumption is a drawback. A manual gearbox would be a welcome addition

Tested	24 Apr 1982
ENGINE	5343cc
Max Power	299hp at 5500rpm
Torque	318lb ft at 3000rpm
Gearing	26.88mph/1000rpm
WARRANTY	12/UL
Insurance Group	9
Automatic	●
5-Speed	N/A
Radio/cassette	●
Sunroof	£855
WEIGHT	3824lb

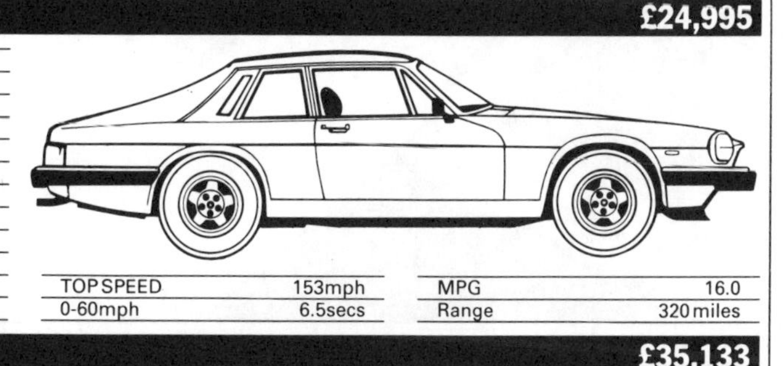

TOP SPEED	153mph	MPG	16.0
0-60mph	6.5secs	Range	320 miles

LAMBORGHINI JALPA — £35,133

Based on the design of the old Urraco Silhouette, the Jalpa is the 'baby' in the Lamborghini family. V8 3.5-litre engine is mounted transversely, driving the rear wheels. Weber carburettors are used, not fuel injection. Body is unitary construction, rather than tubular as with the Countach. Outstanding handling, with braking to match. Strictly a two-seater

Tested	N/A
ENGINE	3485cc
Max Power	255bhp at 7000rpm
Torque	231lb ft at 3500rpm
Gearing	20.8mph/1000rpm
WARRANTY	12/UL
Insurance Group	9
Automatic	N/A
5-Speed	●
Radio/cassette	●
Sunroof	Targa top
WEIGHT	3315lb

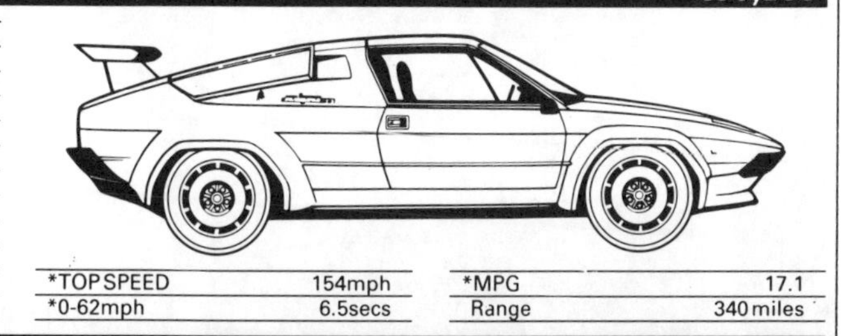

*TOP SPEED	154mph	*MPG	17.1
*0-62mph	6.5secs	Range	340 miles

LOTUS ESPRIT TURBO — £23,440

If one can have such a thing, the Lotus is the efficient sub-supercar, providing near-140mph performance from one of the few acceptable turbo engines; excellent response and wide spread of smooth power belies 2.2-litre origin. Low weight and drag of Guigiaro body helps relatively good fuel consumption. Superb handling; poor vision

Tested	19 Dec 1984
ENGINE	2174cc
Max Power	210bhp at 6250rpm
Torque	200lb ft at 4500rpm
Gearing	22.7mph/1000rpm
WARRANTY	12/UL, 8 anti-rust
Insuranceegroup	9
Automatic	N/A
5-Speed	●
Radio/cassette	£290
Sunroof	£550
WEIGHT	2673lb

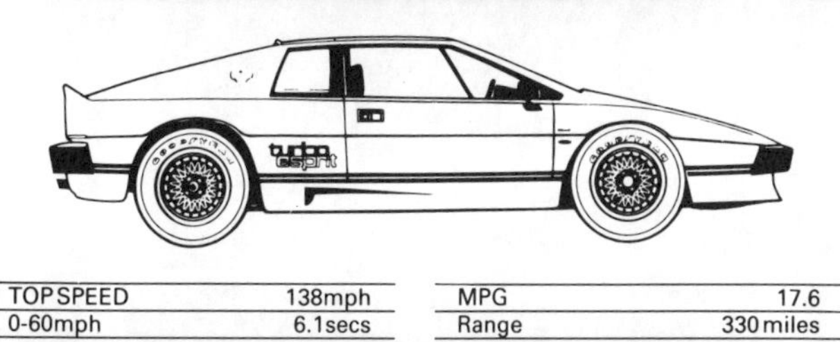

TOP SPEED	138mph	MPG	17.6
0-60mph	6.1secs	Range	330 miles

PORSCHE 911 CARRERA COUPE SE — £36,676

The flat-six, rear engine design of the Carrera has now become a classic. Performance, provided by the 231bhp, 3.2-litre, air-cooled engine, is outstanding. Ultimate handling, especially in the wet, can be a test for the inexperienced driver. Ride quality good, and brakes well able to handle the car's speed. May be called a 2+2, but space is strictly limited

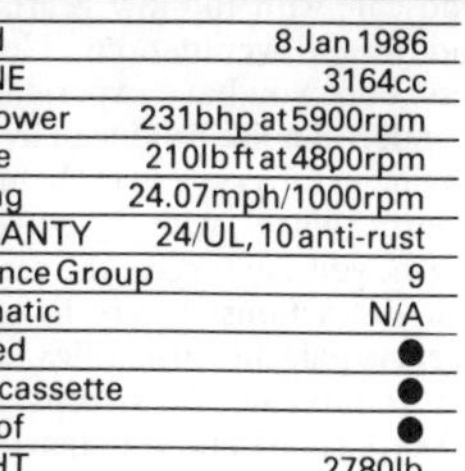

Tested	8 Jan 1986
ENGINE	3164cc
Max Power	231bhp at 5900rpm
Torque	210lb ft at 4800rpm
Gearing	24.07mph/1000rpm
WARRANTY	24/UL, 10 anti-rust
Insurance Group	9
Automatic	N/A
5-Speed	●
Radio/cassette	●
Sunroof	●
WEIGHT	2780lb

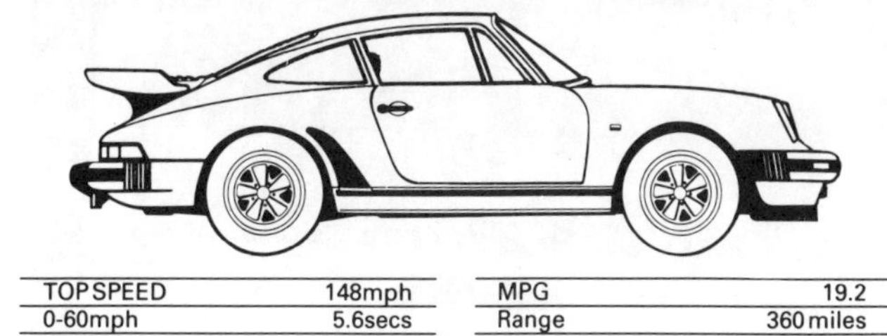

TOP SPEED	148mph	MPG	19.2
0-60mph	5.6secs	Range	360 miles

● Standard N/A Not applicable DO Dealer Option * Manufacturer's figures

Rear seat *leg and headroom is suitable only for small adults or as additional luggage space*

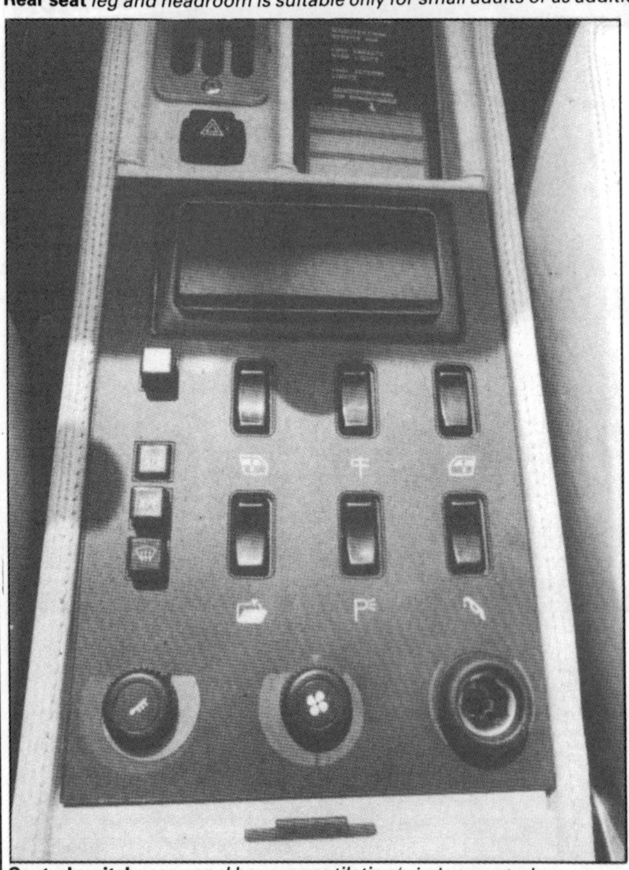
Central switchgear *panel houses ventilation/window controls*

SAFETY

Much of the safety in cars like the Ferrari Mondial starts with the chassis. Give a car of this performance indifferent handling and roadholding and insufficient braking and you have a recipe for disaster. The Mondial's chassis, braking and aerodynamics make it an inherently very stable and safe car. The tubular steel chassis provides immense strength combined with carefully calculated deformation characteristics.

VERDICT

Among the supercars the name Ferrari and the prancing horse badge still reign supreme, and even the relatively 'unsporting' 2+2 Mondial will turn heads wherever it goes. It is a car with few rivals, perhaps the closest being the Porsche 928S, which provides rather better accommodation and performance in a similar bracket to the Mondial.

Yet looked at in some lights, it is an odd car, with too-low gearing, and indifferent ventilation. Nevertheless, once you have experienced the wonderful noise produced by the V8 engine in full cry, and sat behind that steering wheel, with the power surging in, you can forgive the car for the few detractions it may have. They seem to pale into insignificance as the rev counter needle sweeps past the 7000rpm mark and you slam that gear lever through the gate.

It is a different world of motoring.

BRUNDLE'S DREAM

Tyrrell F1 star Martin Brundle always wanted a Ferrari. Recently he bought a 328 GTS, handing over a bundle of money for a car he'd never ever driven. Was he disappointed? Read on . . .

OK, so I'm a Grand Prix driver and I'm offered a discount when buying a new car . . . but isn't everyone these days? "He'll only say nice things, especially as he's writing about a Ferrari . . ." I hear more cynical readers muttering. Let me assure you of two relevant facts. Brought up in a family with strong motortrade ties, I've always been interested in the industry and its latest products – the lean-burn Escorts and revamped Pandas, not just its up-market masterpieces – and before parting with over £30,000 I've got to want that car for many more reasons than just a good deal. This is how owning a dream machine from Europe's most famous manufacturer really is . . .

I collected my Ferrari from Maranello Sales in Egham. A very pleasant lunch with directors Konig and Maingot was followed by my parting with a frighteningly large cheque – the price had gone up by £2700 since placing the order! The organisation and the car itself simply oozed Enthusiasm, but my 328 could have been cleaner both inside and out.

My first few miles with the Ferrari were, not surprisingly, very impressive – I'd never driven one before. As you sit firmly, surrounded by yards of cream leather, looking out across the brilliant red front wings bulging up over those fat Goodyear NCTs, the anticipation is electric. The click of the gearbox gate, the wail of the transverse V8 – sheer joy. However, a few miles on and I realise that the ventilation needs some fine-tuning and the trim rattles a bit. My initial assessment of the interior quality diminishes, but still running-in carefully, I'm falling in love with that engine and drive train.

The Ferrari begs to be driven hard and in the short time I've had the car it's been a struggle not to become an absolute hooligan on the roads. That 3.2-litre four-cam V8 with its 32 valves revs so very freely and, once warm, the gear lever can be thrust across the bright metal gate rapidly and reliably. The dog-leg first gear isn't a problem, though you can't select second for a mile or two – just go up to third. The engine has so much torque you can start off in second anyway, and run from 50 to 150 mph in fifth . . .

Overtaking is mostly carried out by leaving the box in fifth and tickling the throttle. The Ferrari is supremely easy to drive in that respect, but beware of poor road surfaces, cats eyes and even white lines, as all three will have the car changing direction, especially under braking.

This low red blur is a semi-racing car, a fact of which you are reminded as the steering column jumps up and down over bad surfaces and the front valance bottoms out against the tarmac occasionally. I specified the optional air-conditioning, as without it the standard ventilation system leaves you gasping for air.

Getting in and out of the Ferrari calls for some acrobatics, but once seated, I find the all-round visibility much better than I'd expected. I cannot get to like the straight-arm driving position, with the pedals far too close to the seat, and you should forget any notions of hi-fi music, as once on the move the noise of the engine dominates totally. But that wailing engine makes up for stereo shortfall – it is truly glorious . . . for a while. After 150 miles of non-stop motoring, however, I'm pleased to take a break from the high-decibel environment. The ride is surprisingly supple for such a car and is really rather impressive.

The handling is always interesting. Fantastic traction, surefooted grip, with mild oversteer under power. In the wet it understeers around slow corners, and locks front brakes at slow speed, but if you give the throttle a squeeze, you can produce very predictable power slides. You must drive the Ferrari at speed with maximum concentration and respect, because otherwise it will bite back . . . you're always travelling 30 mph faster than you think!

A two-seater Ferrari has to be a second car. It lacks space. If you buy some suitably sized luggage and work hard at the packing the boot space behind the engine will hold a reasonable amount, but the nose is a non-starter as far as cargo goes. It's hard to keep all that red paintwork gleaming as there are so many nooks and crannies where dirt collects and those fins in the cooling intakes are a problem.

Biggest problems of life with the Ferrari are the wind noise – which reduces dramatically when the headlamps are up! – trim rattles and the fact that it attracts so much attention. At first the attention-getting is a novelty, but it soon becomes a pain.

If the Japanese can produce a £10,000 coupé that's packed full of equipment and never breaks down, then a car two or three times that price should never be less than perfect. Strangely, cars at this price invariably are less than perfect, and even more strangely enthusiasts the world over seem to accept – even expect – it. I suppose it's a bit like saying "the MacWillerrari F1 team spends £15 million per year, so how can their car ever not finish a GP?"

At its first service the Ferrari's job list will read "make central locking work (there are only two doors after all!), cure water leaking into interior, cure excessive wind noise (rubber seal on offside A-pillar is an inch too short and moves up/down channel), rid interior of numerous rattles".

Mechanically the 328 is perfection and it returns 20 mpg which is surprisingly good. A £37,000 car should never fail, but if you want a hand-built supercar in these high-volume robot-orientated times, you have to expect some problems.

The Ferrari 328 GTS is, to my eyes, the most beautiful car on the road. It's not the best car I've ever owned – that was a Mercedes 190 E 2.3-16 – but it is a thrill to drive fast on open roads for shortish bursts of adrenalin-pumping entertainment. Maybe it is little more than an executive toy, but when you've decided that life is all too short, order one!

PHOTOGRAPHS BY TIM WREN

FERRARI'S PRETTIEST, THE 328GTS, IS NOW MORE POWERFUL, MORE TRACTABLE, MORE CHUCKABLE. STEVE CROPLEY COMPARES IT WITH HIS OWN R-REG 308GTB, THE 328'S ELDER BROTHER

ROSSO CORSA

AS SOON AS IT WAS BEYOND DOUBT that the Ferrari and I were going to leave the road at speed, several interesting questions began to form inside my head. This may not seem a conventional response, but my foot was already clamped on the brake pedal, the speed was still hopelessly high and whatever actions could be taken had been taken. There seemed to be plenty of time to consider each question thoroughly, between the instant of awful realisation and, a couple of seconds later, the moment we were launched off the road.

How could I have been fool enough to allow this to happen? That was a good one. I mean, I *knew* this road. I'd driven down it often enough before, including just a few minutes earlier in a 1986 Ferrari 328GTS, today's other test car. And at much the same vigorous speed. Yet, here I was, about to wipe the nose off a 1976 Ferrari 308GTB.

Would it hurt much, I wondered? My only other road accident had been about 15years ago and the memory of the pain had faded, but there was no room for optimism.

Another thing: what would I put in my garage now? This potential pile of oily wreckage was my own car, the one high performance machine with which I had become truly familiar because we had done 5000miles – all of them recreational – in the space of a year. It was chilling to realise that here, at this very kink in the road, our happy association was to end.

I didn't bother to speculate on why the 308 hadn't reached down with its Michelin XWX tyres, which were all the rage 10years ago, and gripped the road as well as the Goodyear NCTs of the 328 had done a moment or two earlier. This inadequacy had been clear from the moment I'd gone for the brakes and first felt their, and then the tyres', comparative lack of bite. The point became even clearer when tyre smoke from the locked front wheels began to curl up over the bonnet. Here, in the first minutes of our serious driving phase, was the most graphic possible demonstration of what we had expected to take two days to clarify – the difference 10years of engineering development puts between two cars whose bones and innards are essentially the same.

The fact that the 308 was not smashed to pieces by the rocks that stud the outside of every interesting B-road bend in Wales was no testament to my driving skill. It had more to do with a similar lack of driving skill of the two dozen tourists who had been there before me and, at the expense of their own Cortina and Cavalier front ends, had pushed all the rocks out of the way. They had even worn a usable little track in the hump which should have been a foot high on the outside of the corner. The Ferrari picked the gap neatly, bumped down a steep but grassy slope and came to a stop, without any damage beyond a paint chip, the size of half your thumbnail, off the underside of the front spoiler. It was a miraculous escape which I am ashamed to say was not earned by the exercise of any skill on my part; but I take comfort from the fact that there are now lurid skid marks on the road which will serve to warn others for the rest of the summer.

THE 308 HAS BEEN FERRARI'S BEST combination of road ability and inspiration. With the improvements built into its successor, the 328, and that model's appearance, broadly speaking, just as the wonderful Boxer was replaced by the less than wonderful Testarossa, the 'little' two-seater's importance to Ferrari enthusiasts who really use and drive their cars, has grown acutely. Today, the Testarossa is just too big, wide and expensive, and it isn't the most arresting car in the world, either. The front-engined 412 is overweight and overshadowed by many a Benz coupe, as well as being rivalled by an Aston. The Mondial, despite worthy qualities, has always been just too sensible and just not pretty enough to sell well. Thus, the 308/328 has become the most-built Ferrari in history, with sales homing in on 10,000 since it appeared at the Paris Motor Show of October 1975.

Performance gap is considerable between cars with 270 and 'real' 220bhp, but glass' 308 makes up some headway by being at least 300lb the lighter. Handling of both cars is exemplary but greater grip (and braking bite!) of far superior NCT tyres give big advantage in feel and steering sensitivity to 1986 car

ROSSO CORSA

The thing the 308/328 lacks, of course, is a 12-cylinder engine. I also had misgivings, before I became so familiar with my own car, about the fact that the V8 engine was mounted across the chassis, ahead of the rear wheels, instead of north-south like all self-respecting full-size exotics (except the Lamborghini Miura) had theirs. It seemed a little bit too much like kit car practice to me, where somebody takes an old transverse Maxi power pack, and stuffs it into a backyard frame of waterpipes and garden furniture.

Yet the two-seater V8 is all Ferrari, and all sophisticated design. Its transverse power pack, with the gearbox below and behind, allows it to sit very low, even among Esprits and Urracos, and to be quite short. And the sophistication of the on-paper engineering is translated into practice. The car is based on an extremely tough tubular space-frame with unstressed panels clad to it. The suspension is by wide-based unequal-length wishbones with coil springs and Koni dampers at both ends, plus heavy duty anti-roll bars front and rear. The brakes are all-disc, of course, and the steering is an unassisted rack and pinion system. You'd expect no less from a Ferrari.

When the Ferrari 308 made its debut, 11years ago this October, the Dino 246GT had been a year dead and was much missed. That car had been a fairly cheap, charismatic two-seater that put a lot of Ferrari drivers on the road who wouldn't have made it in an old-style V12. The 246 replacement, the 308GT4 which people called the 'Dino V8' was a Bertone-bodied two-plus-two that was considered plain, and which had a reputation for twitchy handling. But its engine met with most people's approval.

What was needed was a two-seater. The GTB combined the best elements of 246 and Boxer shapes in a shorter, lighter body than the GT4's. And it was immediately successful, so much so that it grew rapidly into the foundation model for the firm. The first 18months' production of 308GTBs had bodies made predominantly of glassfibre, but in mid-1977, a switch was made to all-metal construction.

Nobody at Maranello these days can provide an all-embracing reason for the change. In reality, there were three of equal weight. First, Ferrari wanted to launch the GTS, a Targa-topped version of the GTB which they insisted on referring to as 'Spyder', as if it were a fully open model. Second, the glass' bodies, made at the Scaglietti body works, were too expensive to be economic; and third, Ferrari's salesmen just couldn't get it into the heads of the faithful that plastic was a fit material for a Ferrari. It's ironic that these days the glassfibre models are the ones in demand because they're lighter (by at least 250lb), they are unaffected by the rust that by now has ravaged just about every pre-'80 metal-bodied V6 or V8 Ferrari, and because their carburettor-fed engines are completely unfettered by power-sapping exhaust

308 has stark cockpit (top left). Old Momo wheel gives clue to age. Carb engine (above) is said to yield 255bhp; It is more like 220 in reality. Boot is the same zip-up compartment as today's car; holds a surprising amount but it's best if the bags are soft. Heat can be a problem. Dash (left) is overstuffed with dials, warning lights; console furniture (far left) is from 'metal gadgets' era, but easy to use

328 cockpit is similar to 308's, but everything has been modernised. Thicker-rim wheel is nicer to hold, gearlever is long to give better leverage. Clutch lighter, too. Centrally positioned dials ease clutter; plastic switches are a failure. New seat facings (bottom right) go on same frames but bulk of extra padding restricts room. New 3.2 engine (below) pumps out 270bhp, with lots of torque

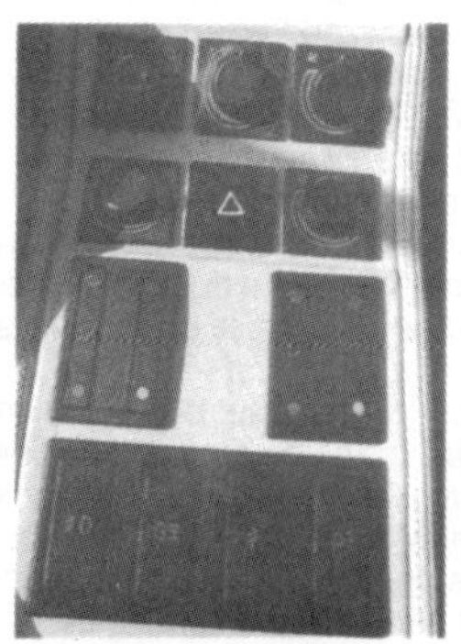

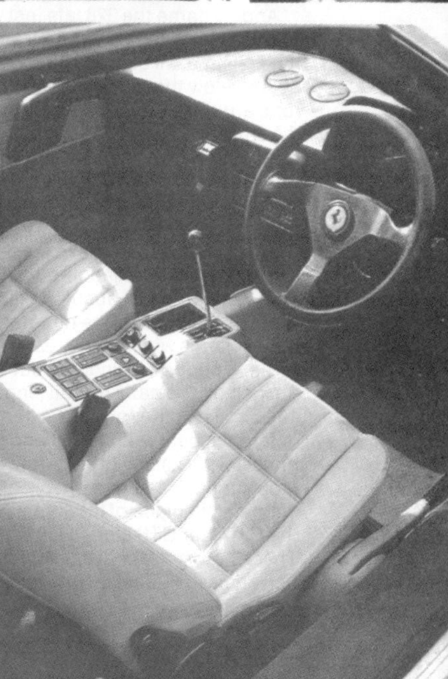

emissions equipment.

After the change to metal bodies, the 308 went into decline. It grew heavier, though at first it kept its carburettor engine's healthy quoted rating of 255bhp. The unit was powerful and very responsive, but it wasn't as strong as the figures suggested. Those were the days when Ferrari could still quote power figures by thinking of a nicely rounded number; it would have been more honest to claim 220bhp or 230.

After the 255bhp engine came a tame 214bhp unit early in 1980, as the world's emissions regulations bit heavily and Ferrari, in pursuit of export sales, had to fit Bosch K-Jetronic fuel injection to the two valves/cylinder V8. The 3.0litre quad-cam V8 engine had been unchanged in major dimension or specification since its debut in the GT4 in 1974. That GTBi, the injection model, made a slow car. It couldn't even crack 140mph unless conditions were ideal. Its predecessors had knocked on the door of 150.

By the end of '81, Ferrari had produced four-valve heads for the 3.0litre V8, and put them in cars called the 308GTB/GTS Quattrovalvole. On a new, honest, rating system the engine gave 240bhp at 7000rpm and 192lb ft of torque at 5000rpm. Though by that stage the car was 300lb heavier than its glass' bodied ancestor, it was a real performance car again. It could exceed 150mph, and people were often beating six seconds for 0-60mph runs.

But by that time, continuous development of the Porsche 911 – ever the middle-sized Ferrari's natural enemy – had produced a 3.2litre car which was just as fast as the best-driven Ferrari and which, on most export markets and especially in Germany, was considerably cheaper, better-braked, more durably built and finished and, because of Porsche's exceptional attention to testing detail, could better stand sustained maximum effort use than any other exotic car ever built. Ferrari, led by Fiat people who knew that product excellence, not badges on noses, keeps cars selling, knew it had to do something extra.

And so came the 328 late last year. The car was altered in appearance for the first time, all of the major changes being below the black belt-line that runs around every 308/328. That gave the shape more bulk down near the road, and took away the needle nose. But it improved its frontal aspect by making it appear to hug the ground more securely.

The new bumpers make solid sense. The 308 has laboured along for nine years with just about the flimsiest protection possible; the new car has bulkier items which add greatly to an owner's peace of mind, but put barely an inch on the car's overall length of 167.5in. The other obvious external difference between the 328 and its ancestors is its use of elegantly redesigned five-spoke wheels of 16in diameter, in place of the 14in hoops the GT4 and GTB started life with. These now run Goodyear NCT tyres. Ferrari used Michelin TRXs for a few years before they settled on the V-rated Goodyears in the 205/55 size for the seven-inch wide front wheels, and 225/50s on the eight-inch rears.

Inside, the 328 has been thoroughly rejigged, to include the pizzazz today's buyers demand. Ferraris up to the Boxer used to have simple, even classical, interior layouts, but the 328 has more dials and switches, better looking dash and seat trims (on the same basic frames), a kind of half-console to carry three ancillary dials, the optional radio (or else an unsightly patch of black vinyl) and a couple of Fiat air vents. Ventilation and window controls are grouped, as before, on the centre console beneath the driver's left elbow, as they always were, but they're plastic 'touch' switches now instead of the metal toggle type which ruled for 10years. They're less clearly labelled and harder to use quickly. The same goes for the door catches, which are not as elegant as they were on the original car. Perhaps the parts bins from which those items came, just emptied out. More likely, there was something wrong with them from the Swiss or Australian or California design rules point of view. But, for people, they were fine.

ROSSO CORSA

The 328's new interior does help justify the basic price of nearly £36,000 better than the original could ever do these days. The bucket seats are a little more thickly padded and a little higher set, the gearlever is longer to give better leverage, the pretty Momo steering wheel's rim is thicker and thus easier to grasp firmly in parking manoeuvres (which still need muscle) and the clutch is much lighter. The main drawbacks seem to be the switchgear and the fact that the front wheel arches have been expanded inwards to accommodate the wider, lower profile rubber of modern times. This severely cramps the pedal room. It's positively easy now for drivers with large feet to go for the brake and include the accelerator.

But the key 328 changes are to the engine. It is both bored and stroked to 3185cc (formerly 2927cc). New pistons incorporate 'squish' theory design, giving better cylinder filling and mixing. The compression ratio is raised from 8.8 to 9.8 to one. The car uses Marelli Microplex electronic ignition, a better system than the former Digiplex system. The oil radiators are redesigned, and myriad improvements aim for the kind of high performance durability that Porsche achieve.

The internal ratios of the five-speed gearbox, mounted below the transverse engine and just ahead of the final drive, are as they were from Day One of the GT4 in 1974, but the overall gearing is altered again in the 328, this time from the 308 Qv's 20.5mph/1000rpm to 20.9mph/1000rpm. The old GT4 had about 22.5mph, the first 308 had 21.5, the injection two-valve 308 went down into the 19s (as Maranello attempted to cover its lack of power), and last year's 308 Qv was back at 20.5mph.

Yet what strikes you when you bring together the first 308 and the latest 328 Ferraris is their string of similarities, even after four distinct model changes. The shapes are so nearly the same. The cars are card-carrying members of a super-exclusive club comprising the purist sports cars that are not an inch larger than they need be. Though infrequently mentioned, that is such an important factor for cars which aim to be enjoyable, day-to-day, on the roads of Britain. Horsepower is no substitute for practical compactness.

The group embraces the non-Turbo Porsche 911 (the widest-wheeled versions are too wide for true manoeuvrability), and would have included the Maserati Merak if it were still alive. The Lamborghini Jalpa and Lotus Esprit don't make it (also too wide) and the Countach/Testarossa brigade wouldn't get a look in.

The 328 fits a pair of people beautifully closely, but there is a price. The seats aren't up to much. The lower part holds your nether bits well enough, but the backrest doesn't have any side support to speak of. If you're of much height at all, your knees rise high either side of the low, raked steering wheel and the gearlever grows out of its gleaming gate close against your left thigh.

FROM THE INSTANT YOU TWIST THE key, you're aware that the 328 is all Ferrari. The starter motor emits the high whine that has become so characteristic of a Maranello engine. It divides the breed irrevocably from the ruh-ruh-ruh of the common herd. The engine first catches unevenly on several cylinders, rather than bursting into life on eight.

No matter how many times you hear it, you are never quite prepared for the excitement of a Ferrari engine note – six, eight or twelve. When you start the 328 you create a note which speaks volumes for the car's potential and its breeding – and which stresses the straightforward relationship between this actual car and some of the greatest race machinery in the world.

Once carefully warmed, the 328 is equipped to go like the clappers. Despite the lingering assumption that this is the 'little' Ferrari, with its new power transfusion the car has a power/weight ratio of just 10.1lb/bhp which is remarkably high for a 3.2litre car. Even the mighty Testarossa isn't so much more powerful at 8.5lb/bhp – and a quickish saloon such as the Golf GTi, must shift 18.5lb with every one of its horsepower. No figures illustrate better than these the differences between supercars and fast saloons.

The Ferrari 328 is now a genuine 160mph motor car. It will pull that speed on any old stretch of unrestricted autobahn. In fact, it will pull its maximum engine speed of 7700rpm in top with surprising ease, an engine speed which works out at just 161 in reality and around 170mph on the relatively accurate speedometer.

But despite the exalted top speed, this car is undergeared, if anything. It's so strong in the mid ranges that you can change from *any* gear to the next gear at 4000rpm, no matter what the topography or driving conditions, and still be going very quickly indeed.

Off the mark, the car chirps its wheels only if the clutch is dropped with more than 5000rpm showing. The limited slip differential and the rearward weight transfer onto the driven wheels keep traction very well. First gear rams the car just beyond 40mph, second, selected with rather a slow, forward dog-leg movement, takes it to 63mph, third runs out just short of 90mph and fourth is good for 119mph. If there's a criticism of the ratio spread it is that second, third and fourth are all rather short. This engine has so much mid-range urge now that you can easily miss second altogether, and go nearly as quickly. And there is quite a gap between fourth and top, as there always has been in transverse V8-engined Ferrari cars.

But the shortish intermediates do make for terrific lower-end acceleration times. The 328 will sprint to 60mph in just 5.8sec, which used to be a good time for a Countach until the latest, 475bhp version came along. And even though third won't pull to 100mph – and three gearchanges are therefore included in the time – the car will just beat 14sec to 100mph, which is Boxer territory, too.

Even when you select top to venture beyond 120mph, the push in the back is still strong and smooth. Not until 140mph does the car's acceleration taper off. The 328 has as much shove at that speed as the old 308 has at about 110, notwithstanding the older car's light weight and the good throttle response of the carburettor engine. The 308 is brisk enough in getting to 60mph from standstill in 7.1sec, and it will still pull a genuine 145mph on a still day – but that's a full performance class below this 1986 V8. In the 500miles we put under the 328's wheels, we never had occasion to wish for more power, or for more flexibility.

Its curious Achilles heel is noise. The engine is fairly quiet when being driven easily, with change points below 5000rpm, but it abruptly begins to shout beyond that rev range. At full noise, it is very loud indeed – obtrusively so – which is a great surprise in a car whose clutch and gearchange have been lightened, and whose interior appointments have been so carefully improved all in the cause of habitability. A Porsche is more refined; even my own '76 model has less of the harsh, ringing noise that comes from the top of the engine under load. You can tell yourself that these sounds are music to the ears of the purist, but I'll guarantee that you won't be saying that after 100miles at really high speed.

On the handling front, the 328 is very much reminiscent of its ancestor. It corners neutrally, tending to understeer a little and display hardly any body roll at all as the speeds rise. In ordinary circumstances on dry roads, you will not break the tail loose, even by stamping on the brakes in bends. The car just goes around. It has the usual useful facility of tightening its line neatly when you throttle off in mid-bend, something you'd expect. But, like all mid-engined cars, it gives little idea, before the event, about its breakaway characteristics which are bound to be abrupt, if only because the car will be going very, very hard when it happens. But there's certainly no Porsche-style treachery on the over-run. These comments apply pretty well to the old 308, which simply generates bigger slip angles at lower speeds on its tall old Michelin XWXs. But it still grips very well, and it takes a bold driver to push it to breakaway (at the front) in the dry. In the wet, it will power-slide quite usefully provided you use plenty of power.

The brakes, you know about. The 328's are absolutely first rate, with terrific initial bite even though they're not too light. The 328's four ventilated discs are aided in their work by the considerable adhesion which the tyres have, which postpones lock-up quite a long way beyond the point at which an XWX-shod 308 suffers – as I found – from total loss of adhesion.

On the ride front the cars, 10years apart, are almost equal. If anything, the old car with its flexible tyre sidewalls and squirming tread, has a minor advantage. It's certainly quieter over the bumps, because the sounds echo less around its glass' body than they do inside the 328's galvanised tin one. But both cars lack the suspension travel it takes to cope with really big holes in the road, and can occasionally be comprehensively upset because of it.

What the 308's metamorphosis into the 328 shows, above all, are the influences that have affected all car designers in the past decade. This car, more sensible than most because it was born of the first Arab oil embargos, has become steadily more practical, more livable, more capable of complying with the world's cocktail of design legislation. It has become cleaner of exhaust, better to crash in, better to park in supermarket car parks, less likely to break down. And with its extra speed has come a lot of extra economy. Where the old carburettor car can manage 19mpg on a couple of hundred miles of Sunday morning sprinting, the 328 returns anything up to 23mpg, and probably goes harder on the journey. Driven gently, it can do 25mpg.

Though economy hardly seems the point in such a car, it is, like better bumpers, a key factor when you come around to owning one. The fact that it is accompanied by performance that would have distinguished a Countach just a couple of years ago, shows the giant strides that car designers have made in the past few years – without always seeming to.

The Ferrari 308 and 328 are very special cars to me. Their inspired packaging makes them a joy to use; their handling and performance are ever stimulating; their looks, I think, are a supreme Pininfarina achievement – the more so because the 308 and 328 eschew the awful 'millionaire' label of bigger league exotics, where prices matter more than capabilities. How extremely glad I am not to have reduced the world's Ferrari V8 population by one.

Exterior differences between cars are mainly below belt line. Greater bulk of 328 front section gives car more modern look, as well as better bump protection. Tiny bumpers of 308 (middle) contrast with 328 (bottom) which have spring-back facility, yet add only about an inch to car's length. Rear side slats of 328 restrict vision

ROSSO CORSA

turbo
MASTER BLASTER

ou thought a Ferrari 328 GTS was quick, take a ride
Hans-Jürgen Tücherer in the 450 bhp twin-turbo GTS
's on offer from German tuner Willy Koenig

Main pic and above: Big red booster – Koenig's twin turbo 328 GTS; conversion alone costs almost as much as the car. Below: New, deeper front spoiler keeps the nose down at 190 mph

We didn't reach 200 mph. The arrow-straight section of autobahn to the south of Munich wasn't quite empty enough to allow us to squeeze the last drops of performance from Willy Koenig's remarkable twin turbo Ferrari 328.

In the end, we had to be content with 190.2 mph – 306 kph – though with thick, towering pines straddling the road, the accompanying blur in the corner of my eye gave the impression we were travelling at twice that speed.

Yes, it's quick, very quick. Blast off from standstill and the red monster will flash to 62 mph (100 kph) in a mind-boggling 4.5 sec. Compare that to the standard 328's time of around six seconds and you can see just how quick it is.

Koenig has a fascination for improving the performance of the Prancing Horse – taking the finest piece of Italian exotica and adding that final touch of German engineering. Before the 328, the former racing driver had turned the already rapid Testarossa into a 710 bhp turbo rocket.

He carries out his conversions from a factory on the outskirts of Munich, working with turbo engineer Frans Albert and body specialist Vittorio Strosek. It was Strosek who produced the striking Porsche 944 Turbo featured in last week's *Motor*.

Taking the standard 3.2-litre, 270 bhp Ferrari V8, Koenig rebuilds it from the sump upwards. He reprofiles the shape of the combustion chamber, adds water injection to cool the charge of air, fits special turbo pistons and enlarges the water and oil-cooling systems.

Each bank of cylinders is then equipped with Rajay turbocharger and wastegate which provide a boost pressure of 0.7 bar. The compression ratio has already been reduced from the standard engine's 9.8:1 to 8.0:1. The Bosch K-Jetronic injection system and Marelli's most sophisticated Microplex engine management system remain unaltered.

The result is a solid 450 bhp at 7000 rpm which puts it alongside the 455 bhp of Lamborghini's Countach QV and way ahead of the Porsche 911 Turbo's 300 bhp and even the Testarossa with its 390 bhp.

To make real use of this extra power Koenig has lengthened the final drive ratio by 13 per cent. Had it not been changed it would have been impossible to contain the wheelspin and 190 mph top speeds would have been out of the question. But it also helps keep the fuel bills down; during our test the fuel consumption seldom dropped below 15 mpg.

While few people would dare to criticise the handling of the standard Ferrari 328, Koenig felt that with 450 bhp on tap, a few suspension modifications would help the driver keep the car heading roughly in the right direction on a twisty road.

So the Turbo gets special dampers, stiffer and shorter springs which lower the car by 3 cm, and beefier brakes. Then at each corner go 8½ in three-piece alloy rims shod with 285/50 VR 15 rubber.

To accommodate the extra-width wheels, Koenig called on Strosek to come up with a number of body modifications – half through necessity, half through the need for the car to look different. On went the big front spoiler with its air intakes to help cool the front discs; the front wings were widened; and side skirts with large air intakes were grafted on to the wider rear wings.

Where Strosek went a little over the top was with his giant wing mounted on top of the rear deck. The Arabs no doubt adore it but it owes little to subtlety. Koenig likes it because he believes that without it, life would be quite interesting at 190 mph.

At the rear, there's also a new rear skirt from which protrude the four enormous exhaust pipes, while new door mirrors complete

Above: Strosek-designed body add-ons include side skirts and bulbous rear arches. Below: Turbocharging takes power up from 270 to 450 bhp

Below: Mickey Mouse add-on – the flashy rear wing is big on image, short on style

the body styling package.

Now to business. Turn the key and the big V8 fires up instantly and idles with all the clatter and whir that is the hallmark of Ferrari's 32-valve quad cam all-alloy engine. At low revs the sound is more hollow, slightly harsh and thrummy than deep, bellowy V8 – it's only when you squeeze the throttle that the note becomes more purposeful.

Low speed flexibility is good but it's not until the tachometer needle touches the 3000 rpm mark does the power really flow. Then it's like flicking a switch that releases a flood of raw, neck-snapping power. It's at its most exciting darting from one corner to the next on a twisty, winding cross-country road or car-hopping on single-lane roads when you see the gap in the traffic ahead and squirt past with total ease.

Ferrari's metal-gate gearchange remains, punishing the casual hand with obstinate baulking. But guide the slim chrome lever with deliberation and the change is smooth and precise and complemented by a surprisingly light and well-cushioned clutch.

Despite the car's extra width, Koenig's Ferrari is anything but a handful on the road. Like the standard car, the road manners are impeccable with responses to the helm being quick and accurate. The larger front tyres do increase the steering weight slightly, but in the dry offer sensational grip with the stiffer suspension producing almost no body roll. The ride though is a little firmer with more bumps filtering through the suspension into the cabin.

On the autobahn, the bright red, be-spoilered Ferrari has an amazing effect on other drivers. Without the normal prompting flashes of headlights, they move over as the car fills their rear view mirror. But when you do approach the wayward VW Polo travelling at a steady 60 mph when you're doing 150, the extra braking performance comes into its own, hauling you down with ease.

If the car has one drawback it is cost. Koenig conversions don't come cheap, but at around 200,000 DM – close to £60,000 – it is cheaper than a Porsche 959 or even a secondhand Ferrari GTO. Form a queue here.

Ferrari 328 GTS

Wealth has its rewards

by Jack R. Nerad

PHOTOGRAPHY BY JIM BROWN

In an age of conspicuous consumption, where something more expensive is necessarily better than *any*thing less expensive, there is no better purchase than the Ferrari 328GTS. First of all, there is the price—$62,500—which is high enough to keep out the riff-raff (unless the riff-raff happen to traffic in contraband pharmaceuticals). Even better, the price includes a $1500 "gas guzzler" tax levied by the federal government. Talk about conspicuous consumption. The tax means the well-heeled buyer is ponying up a grand and a half for the *privilege* of paying for more gas than everyone else.

While all this may seem quite negative to those ecologists out there in the audience with Save the Whales stickers plastered to the bumpers of their Saabs, let me assure you that, in the Right Circles, all this is quite marvelous. What good is being rich if commoners with considerably less money can afford to buy the same toys?

There is no doubt that, by all empirical measures, the '86 Chevrolet Corvette, at less than half the price, is at least the Ferrari's equal in performance. The Vette goes from 0-60 in 6.37 sec; the 328GTS in 6.47. The Vette does the quarter mile in 14.85 sec at 91.9 mph; the 328GTS in 14.95 at 95.4 mph. The Vette turns in a 0.92 g skidpad figure; the 328GTS a 0.86 g. The Vette stops from 60 mph in 127 ft; the 328GTS in 152 ft. Thus, with geometric logic, as Captain Queeg once said aboard the *U.S.S. Caine*, the Corvette can be proven to be the equal of the megabuck 328GTS.

What the 328GTS does best is stop people—dead in their tracks

But logic becomes meaningless when discussing Ferraris. Their lure lies in the aura that surrounds them quite as much as the actual hardware itself. Has Chevrolet won eight Formula One titles? Has a Corvette ever won the 24 Hours of LeMans? Did Fangio ever pilot a Chevelle at Daytona?

The fact is that people who don't know anything about cars know about Ferraris. They recognize the name, the panache, the price. A Ferrari, even in the car-crazy climate of L.A., will get you noticed. It's a simple signboard of success, American-style.

But what of the iron itself? Well, frankly, it's excellent. Some Ferraris of the past have put their wealthy owners through contortions of body, will, and pocketbook that no one, not even your most ill-regarded relative, should be forced to bear. They rusted; they overheated; they failed to start. But this 328GTS seems to be—can you believe it?—a practical car. It starts in the morning; the electrics work; it doesn't beat you up with an overly stiff suspension; it even has effective air conditioning.

Under its beautifully sculpted Italian designer skin lies one of the most sophisticated, race-bred collections of pieces available on this orb. The all-independent suspension utilizes unequal-length A-arms, anti-roll bars, and coil springs front and rear. The tuning is remarkably supple, yet the 328GTS sticks.

Our only complaint is the car's tendency to understeer. With the V-8 engine mounted amidships in the Formula One mode for optimum weight distribution, you might expect the 328GTS to be slightly tail-happy, particularly in drop-throttle situations. To curb this tendency for the rear to lead the front end through corners and to keep unsuspecting customers out of the weeds, Ferrari engineers have factored in a bunch of understeer. When pushed, the 328GTS pushes, which came as something of a surprise to one of our staffers in a blind corner at speed. As his car understeered across the centerline, an interloper traveling in the other direction did the same thing. A very expensive rearrangement of sheetmetal was averted only by the Ferrari's quick and precise rack-and-pinion steering.

While understeer is the safest modus operandi for most drivers most of the time, we were actually surprised at the amount in the Ferrari. We had a chance to play hare-and-hound with the 328GTS and a Corvette Roadster and found the Roadster much easier to drive fast because of its neutral steering. We were also surprised at the brake dive engineered in (or more precisely not engineered out of) the suspension. In hard braking, a great deal of weight seemed to rush forward, making it relatively easy to lock up the front brakes. Steering definitely suffers in such situations.

We have no complaint at all about the tires, however. The Goodyear NCT radials (205/55VR16 front and 225/50VR16 rear) proved eminently sticky with good ride and fair noise characteristics.

Some purists claim the only true Ferraris are equipped with 12-cylinder engines, but we think the 328GTS DOHC 32-valve V-8 does just fine, thank you. It might not sing like the 12, but it plays a very pleasant ditty of its own, to the tune of 260 hp at 7000 rpm. The recipient of a bore-and-stroke treatment that has brought its displacement up to 3186 cc and a hike in compression from

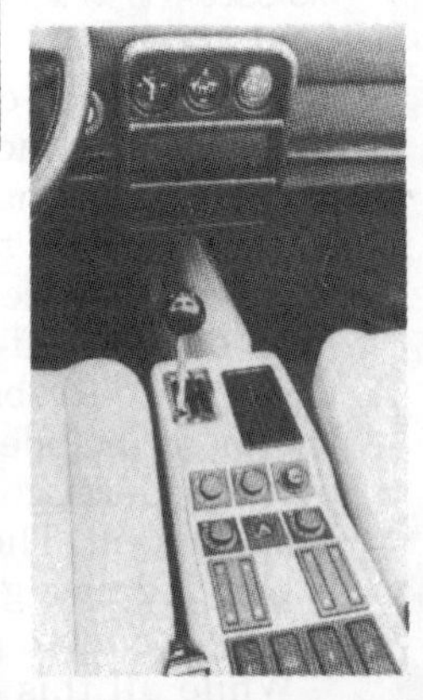

8.6:1 to 9.2:1, the Bosch K-Jetronic fuel-injected engine performs the neat trick of being strong at the low end and peaky at the same time. What this means is the engine is comfortable loafing around town in 5th and becomes an absolute screamer with the pedal to the metal. Go-fast mode definitely involves heavy doses of high rpm, so keep that crankshaft whirring and you'll keep a smile on your face.

To leave the revs up where they belong, you'll be assisted by the 328GTS sweet 5-speed gearbox. It proved a little balky when cold (no doubt a hot-blooded Italian), but properly warmed up the gated shifter was truly satisfying. *We* quickly warmed up to the left-and-down, race-style placement of 1st and found shifting to be a pleasure both in town and over the crooked roads that are our personal favorites.

Inside the cockpit, the Ferrari turned out to be far from the torture rack that some low-slung exotics are. The car stands just 44.1 in. high, yet there is plenty of head and leg room. We might have liked a bit more lateral support from the leather-covered bucket seats, but they're exceedingly comfortable and much easier to get in and out of than the Corvette's heavily bolstered space chairs.

The instrumentation under the traditional Ferrari binnacle is both

The DOHC 32-valve V-8 sings a very pleasant ditty, to the tune of 260 hp

handsome and complete. Some staffers prefer white-on-black rather than orange-on-black markings, but that's a quibble of the most niggling kind. The attractive Momo steering wheel is of appropriate thick-section and richly leather-covered. The fixed steering wheel position is in the Italian/Ralph Kramden style, but once you get in the habit of gripping it at the bottom rather than the top, the bus driver mode becomes rather natural. If you want to go faster—well, lean forward.

Significantly less straightforward were the Ferrari's climate controls. We spent several days in the car and still couldn't quite figure out what all the buttons and knobs on the console did. The main thing, however, was that we did get the air conditioner to blow out copious amounts of cool air, a stunt some earlier Ferraris had difficulty performing.

While we're talking about stunts, we should mention an interesting exercise called "Find the Hidden Door Handle." It was our misfortune to attempt to perform this mission for the first time as we pulled up to a full service gas pump. The attendant had a jolly time watching us trying to extricate ourselves from the red machine before we found the elusive handle tucked behind a door pull which, incidentally, affords you absolutely no leverage in opening the door. Someone should tell the designer about Archimedes.

We must admit the designer got it right on the outside. The new front end cleans up what was already an arresting looking car. The Ferrari is a bit beefy at around 3100 lb, but with 260 hp on tap, we're not going to complain. We do, however, bemoan the rather lengthy 60-0 stopping distance, caused in part by the weight and by the tendency of the fronts to lock. A car with 11.2-in. front and 11-in. rear vented rotor discs should stop better.

What the 328GTS does best, though, is stop people—in their tracks. Drive one and you gain instant star status. Blonde ladies in GTIs cross three lanes of freeway traffic just to see who's behind the wheel. Brunettes in IROCs challenge you to drag races. Not everyone can pull it off, because owning a Ferrari does have its price, in a number of ways. So what? Fame's worth it, and so's the 328GTS. MT

TECH DATA

1986 Ferrari 328GTS

POWERTRAIN

Vehicle configuration	Mid-engine, rear drive
Engine configuration	V-8, DOHC, 4 valves/cylinder
Displacement	3186 cc (194.5 cu in.)
Max. power (SAE net)	260 hp @ 7000 rpm
Max. torque (SAE net)	213 lb-ft @ 5500 rpm
Transmission	5-sp. man.
Final drive ratio	4.06:1

CHASSIS

Suspension, f/r	Independent/independent
Brakes, f/r	Disc/disc
Steering	Rack and pinion
Wheels	16 x 7.0 in. front, 16 x 8.0 in. rear, cast alloy
Tires	205/55VR16 front, 225/50VR16 rear

DIMENSIONS

Wheelbase	2350 mm (92.5 in.)
Overall length	4295 mm (169.1 in.)
Curb weight	1438 kg (3170 lb)
Fuel capacity	70.0 L (18.5 gal)

PERFORMANCE

Acceleration, 0-60	6.47 sec
Standing quarter mile	14.95 sec/95.4 mph
Braking, 60-0	152 ft
Lateral acceleration	0.86 g

BASE PRICE	$62,500
PRICE AS TESTED	$62,500

As fast as it looks, the 328GTS' multi-valve V-8 pumps out 260 hp in a throaty roar—sweet music from amidships.

Updating an Old Friend

Like meeting an old friend after a long absence, an encounter with the Ferrari 328 was a comfortable occasion. You find yourself regarding the car with a relaxed sense of familiarity.

Basically, previous experience with the 308s has taught you the ropes, so it's just a question of picking up the business of Ferrari enjoyment where we left off in 1983 with the last test of the 308GTB Quattrovalvole.

However, when the 328 arrived at our offices late one winter evening, the process of re-acquaintance was a little on the painful side. Memo to my own personal notebook: the central-engined Ferrari is *not* a heavy traffic car. As we joined the rush hour crawl I was instantly reminded just what a fish out of water the 328 was in the bumper-to-bumper evening procession out of the metropolis.

Lowering the driver's side electric window produced a douche of spray into my right ear from the tyres of neighbouring trucks. The clutch is not unduly heavy, but it is certainly heavy enough to make life unpleasant in these claustrophobic circumstances.

The irritating way in which the sharply raked screen reflects the instrument lighting is bad enough on empty roads, added to which the top of the screen's area swept by the single wiper was only fractionally above my eye line. Oh yes, and the wiper itself had emitted a mouse-like squeak. And the steering lock was barely sufficient for manoeuvring in tight corners.

Twenty-four hours later, those trials and tribulations had been blown away, amid the joys of a day's motoring on deserted country roads through north Essex and south Suffolk. The chassis is just fantastic.

All the superlatives have been trotted out time and again, but its impeccable balance, terrific grip and uncanny stability never fail to

A brace of 328s: the Ferrari provided an excuse to inspect David Holland's fine example of its BMW namesake.

arouse the taste buds of even the most jaded motoring palate. And, believe me, I had felt pretty jaded and bruised after my exasperating battle with the rest of the motoring world only the previous evening!

According to factory sources, this is almost certainly the final derivative of the magnificent Pininfarina-bodied two seater coupé which made its bow back in 1975. The transverse mounted V8, enlarged by 200cc, now produces 30bhp more than its 3-litre predecessor, and also develops a worthwhile amount of additional torque which enhances the engine's already impressive flexibility and docile character.

A sleek car in its day, BMW's tall grille contrasts with Ferrari's new bonnet louvres.

The sweet-revving V8 remains mated to its familiar five-speed gearbox, the lever sprouting from its evocative metal gate in the cockpit. Weighted in the second/third plane, the box continues to be adequate, if hardly outstanding by modern standards.

Needing firm and uncompromising handling, the lever moves with a notchy precision — except that dog-leg back into first which has always proved a bit of a trial. Happily, once on the move you can forget about first completely and even from a walking pace the 328 will catapult away when second is engaged.

The braking system remains generally unchanged, with ventilated discs on all four wheels, but the handbrake now works on small rear drums in place of the disc-braked system which was generally regarded as less than effective on the 308. The calipers employed are now the same as those on the Mondial, offering more pad area.

As far as interior trim is concerned, the 328's fascia is now best described as "right-hand drive GTO", the instrumentation being the same as in Maranello's now sold-out range of turbo super-cars.

A triumph of aesthetics — both designs were considered ahead of their period, and neither shows its age.

Snuggling into the cockpit, the driver is faced by a 185mph speedometer and rev counter red-lined at 7700rpm. When working the engine hard the oil pressure remains constant at 85psi with water temperature never climbing above 170 degrees.

The door handles, interior pockets and arm rests are all new, as are the controls for the heating and demisting system, which are still mounted between the seats. Personally, I preferred the old sliding lever system to the current colour-coded, illuminated touch-sensitive controls on the 328, but the new layout is quite logical, even if the 328 demisting system is every bit as slow to produce results as the 308's.

The front and rear bumpers are now colour-coded to match the body's paintwork, the alloy wheels have been re-styled and the cooling vents on top of the front wheel arches have been replaced by additional venting in the centre of the nose section, between the retractable headlights.

Our test car was not fitted with the optional roof spoiler just above the engine bay, there no longer being any choice when it comes to the chin spoiler at the front. Previously, the 308 had been offered with an alternative, deeper spoiler, but the 328's standard kit represents a compromise between the two choices offered on the earlier car.

Living with a Ferrari requires a few days' acclimatisation, and then you suddenly wonder how you ever got along without it. In fact, the 328's driveability, lack of temperament and overall blend of performance and docility tends, by strange paradox, to work against them.

However well one is acquainted with their qualities, there is still a sub-conscious tendency to approach a Ferrari expecting it to be temperamental and slightly difficult to manage on anything but an open road. When you are reminded that they are as tractable and usable as any high performance saloon, you run the danger of comparing them with products of BMW and Mercedes.

But to itemise the awkward aspects of living

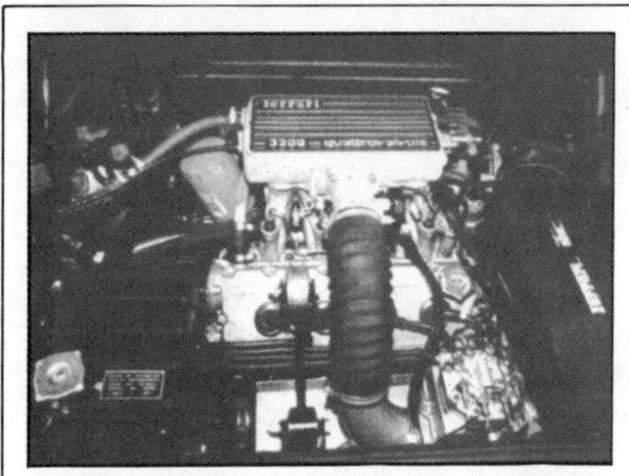

Injected jewel; improved torque and flexibility enhance Ferrari's V8.

Traditional gear-lever is deliberate and evocative Sixties hangover.

with a 328 in the light of how it stacks up against such rivals is totally unfair, albeit highlighting just how well Maranello has done its job in recent years.

Close examination of trim standards, paintwork and general build quality underlines just what a high quality product is on offer. The paint on our test car was of a lustrous quality with no flaws to be seen; similarly the leather-trimmed cockpit had no signs of compromise, botching or shoddy workmanship.

Firing up the transverse V8 from cold is one of the great motoring treats of the decade. A touch on the key and the Bosch K-Jetronic injected jewel bursts into life, ticking over with a gruff exhaust note that belies its smoothness once on the move.

Engaging first gear when the box is cold can be a bit of a pain, but the transmission warms up quite quickly and the whole package has a taut, unified feeling at speed, making the 328 feel smaller than its outward dimensions.

Unquestionably, it is difficult to handle in crowded conditions, reversing into tight spaces being a complex enough business without the added frustration of absurdly small rear view mirrors. The noise level inside is fairly high, a degree of resonance and boom from the neatly packaged V8 proving to be another wearing aspect in slow moving traffic. But at speed on the open road you lose much of it behind you, drowned by the willing wail of Maranello's 270 Prancing Horses.

There is still a reassuring touch of roll when the 328 is cornered hard, sufficient to impart a welcome degree of "feel" to the driver, although it could certainly never be accused of being sloppy by any standards.

As on the 308, I felt the steering a trifle low-geared for my taste, so life can be a little nerve-wracking darting through country lanes. But on more open B or C roads, this brand of Ferrari motoring is nothing less than a supreme joy.

Of course, in terms of pure straight line acceleration this Ferrari is certainly no slouch. It sprints up to 60mph from rest in a shade under six seconds, reaching 100mph in 14.7 sec, by which time it is pulling strongly in fourth gear. A final upchange to fifth at 117mph and the surge of acceleration continues steadily towards its 151mph maximum.

Ferrari's progressive refinement of the V8 two seater coupé has been unrelenting over the past five years, more than compensating for the original loss of performance prompted by the switch from carburetters to fuel injection on the old 16-valve 308 at the turn of the 1980s.

The four-valve (QV) heads redressed the balance even further, but the 3.2-litre model has polished the Maranello veneer to fresh standards of excellence.

It may well be the last of its line, but, unquestionably, it is the best. **AH**

"Nothing less than a supreme joy . . ."

Model: Ferrari 328 GTB.
Maker: Ferrari Automobili, Maranello, Italy.
Type: Two door, two seater coupé.
Engine: Light alloy 90-degree V8, 3185cc (83 × 73.6mm). Twin overhead camshafts, cr 9.8:1. 270bhp at 7000rpm. Bosch K-Jetronic mechanical fuel injection.
Transmission: Rear-wheel drive. Five-speed manual transmission with limited-slip differential.
Suspension: (front) Wishbones and coil spring/dampers and anti-roll bar. (rear) Wishbones and coil spring/dampers and anti-roll bar.
Brakes: Ventilated discs on all four wheels with servo assistance and split circuits. Handbrake operates on rear drum.
Steering: Rack and pinion.
Wheels and tyres: Light alloy 7in rims (front) with 205/55 VR16 Goodyear NCT radials. Light alloy 8in rims (rear) with 225/50 VR16 NCT radials.
Performance: 0-60mph, 5.7 sec; 50-70mph, 5.4 sec; Maximum speed: 151mph.
Economy: Overall, 18.7mpg. Estimated 24mpg (touring).
Price: £38,900.14p basic (tax paid). Extras fitted to test car included metallic paint (£599.25p), air conditioning £1499.99p). Maranello Concessionaires charge £200 for delivery in the UK, plus £20 for plates.
Summary: One of the great, thoroughly usable high performance sports cars of our time. Cumbersome to manoeuvre in traffic, it belongs on the open road where its exquisite road manners can be exploited to the full.

REVIEW

FERRARI MONDIAL 3.2

Ten days with Ferrari's most practical product.

BY DAVID E. DAVIS, JR.

Ann Arbor—I came back from a visit to the Ferrari works in Maranello, stopped off in Hasbrouck Heights, New Jersey—home of Ferrari North America—and picked up a lovely red 3.2 Mondial coupe for the 650-mile drive home to Ann Arbor.

The employee who led me to the car in the parking structure seemed to be innocent of any useful information concerning its care, feeding, or proper operation. So, not yet able to open the luggage compartment, I tossed my duffel into the back seat, and I was off. Fortunately, only a few minutes later, traffic was more or less at a standstill on westbound Interstate 80, and I was able to read the owner's manual as I crept along at a glacial pace in the fast lane.

Knowledgeable *Ferraristi* appreciate the Mondial's role as the "nicest," most accommodating Ferrari of the bunch, and our local Ferrari dealer, Mr. Bob Schneider, of Dearborn Sports Car Exchange, has described the Testarossa as being "every bit as pleasant to drive as a Mondial." Our chief European correspondent, Mr. Kacher, once suggested that we do a story entitled, "Mondial: The Best Ferrari of Them All?" Ferrari's general manager, Ing. Giovanni Battista Razelli, confided that he preferred the Mondial, and that the fastest time for his frequent trips from Modena to Genoa was achieved in a 3.2 Mondial. It is not *fast,* in the sense that a Testarossa is fast, but it is an extremely docile, comfortable Ferrari that encourages its driver to make full use of all the performance it has.

PHOTOGRAPHY BY COLIN CURWOOD

FERRARI MONDIAL 3.2

The Mondial coupe was introduced at the 1980 Geneva Motor Show. In autumn 1982 it received the four-valve (Quattrovalvole) cylinder heads, and in January 1983, it debuted at the Brussels show in its convertible form. The 3.2-liter engine was introduced at the Frankfurt show in 1985. The convertible was originally conceived strictly as an American model, but within a year it was also being offered in European trim. Convertibles account for about three-quarters of U.S. Mondial sales, and Mondials represent maybe twenty percent of Ferrari's 1100-unit annual sales total in the United States. The early 3.0 Mondials were bog-slow, and far less sexy-looking than their contemporaries, the 308s. The car got off to a slow start, and has never really caught up with its market.

Like the 328 models, the Mondial is a mid-engined car, sharing virtually all of the two-seaters' engine, driveline, and suspension components. The wheelbase is 11.8 inches longer to accommodate the rear seats, but it would have to be at least a foot longer still before those rear seats actually became functional. It would be a much more useful car if its engine were mounted in the front, driving the rear wheels. Nonetheless, the rear seat area does provide useful luggage space, and the lengthened wheelbase does improve the Mondial's ride, and seems to make it more predictable in fast transitions. A dab at the brake, a little steering, and the tail seems to know exactly where you'd like it to be.

Twenty-five miles east of the Delaware Water Gap, traffic cleared up and I had gleaned everything I needed to know from the owner's manual, and I started to motor. My Escort's plug didn't fit the Ferrari's lighter, so I was driving naked, but by hooking up with one fast guy or another I was able to cruise across Pennsylvania at seventy-five or eighty most of the time. Twice I blundered into radar traps at those speeds, and both times I was ignored. Was I invisible? Had I inadvertently stumbled upon Lamont Cranston's power to cloud men's minds? Probably not. More likely the policemen were just immersed in their comic books.

Interstate 80 is a lovely road where it snakes through the hills in central and western Pennsylvania, and I really felt that I was getting to know the Mondial. The driver's situation is the best of any of the contemporary Ferraris. The seat can be adjusted to the driver's specific peculiarities without having to make compromises for headroom and legroom—unlike the 328, for instance. When the seat is properly adjusted, the visibility is quite remarkable. I found that I couldn't see the front-end sheetmetal at all, and I liked the feeling that I was flying along, not as a passenger in a machine that surrounded me, but as an integral component of that machine.

The Mondial's steering wheel is mounted at the traditional Italian Trailways bus angle, and I found myself quite comfortable steering the way I've seen Europeans do—pulling down on the steering wheel rim with the hand that's on the inside of the corner, and pushing up with the hand that's on the outside. Jean Lindamood describes this as "the Italian technique of shuffling the wheel through your hands." Works like a charm.

Like the steering wheel angle, Ferrari's shift linkage and shift gate have received more than their share of criticism from America's automotive pundits. A Ferrari is not as effortless to shift as a Toyota or a Honda, but, with practice, it has charms that go beyond mere ease of shifting. It's worth pointing out that if Ferraris were as easy to drive as Hondas and Toyotas, then they'd *all* be bought by the gold chain crowd.

REACTIONS

An everyday Ferrari.

It is axiomatic that an automobile with tremendous character cannot be all things to all people. But does that mean that breadth of appeal is inversely proportionate to strength of character? As a car manages to become more accommodating, does it necessarily become less interesting? I don't think the equation is quite that simple, but there are certainly tradeoffs involved.

Case in point: The Mondial is the most useful, most flexible car in the Ferrari lineup. Compared with the faster and more confining Testarossa, or with the faster and *much* more confining 328GTB/GTS cars, the Mondial is roomy, comfortable, and unchallenging to drive. It is acceptable for both touring and commuting in that strange and wonderful place called the "real world." On balance, the Mondial is a pretty decent automobile, and an incredibly good Ferrari.

But I can't shake the sense that it is also, somehow, a *lesser* Ferrari. Swoopy styling, an intimate cockpit, and snappy acceleration are part of what Ferraris are all about. Making sense as transportation was never part of the mandate. Even if it is for a good cause—civility and accommodation—that the Mondial's lustier qualities are toned down, they are still diminished. Mr. Davis feels comfortable with the tradeoffs; I have a harder time with them. Sure, I like the Mondial, and I respect the job Ferrari did with it. But if I could have only one Ferrari, there's no question it would have only two seats. —Kevin Smith

The Mondial is an everyday car; an everyday Ferrari, if you will. Even so, I couldn't help falling for it during its altogether too brief visit. Part of my attraction was for its redness, part for its lovely lines, part for the fact that it's a Ferrari—a magic name that I don't have the desire to ignore, even in the name of journalistic objectivity. In any case, as Mr. Davis says, the Mondial is so easy to drive, so comfortable to fit into, so inviting, that the rest is mere icing on the cake.

Journalists aren't supposed to be this shallow, but frankly, I was half-hooked when I opened the Mondial's door and the smell of leather (not the coddled smell of garment leather, nor of "Corinthian" leather, but the raw, masculine smell of, ahem, *hides*) almost knocked me down. There is something masculine and tough about all Ferraris. It is that smell. It is the gleaming chrome of the external shift gate. It is the unearthly howl of the engine at 7000 rpm hammering the back of your skull, and the discipline it takes to ratchet the stick cleanly up and down through the gates.

Tough girls like me go crazy for men's cars. And I don't think for a minute that a couple of extra seats in back and a comfortable position behind the wheel make this particular Ferrari any less of a man's car. The Mondial is still demanding, only it's not cramped for space, and it's not as compromising to drive as its two-seater progenitor. Hand me the keys and I'll head for tomorrow. —Jean Lindamood

David E. Davis, Jr., and Kevin Smith may tell the world that, as Ferraris go, this one's easy to drive. Let's put that in perspective. I've seen the two of them curse and struggle to find reverse. I've seen them perform all manner of slick two-steps and *still* miss shifts. I've seen both of them saw at the wheel in the parking lot, making two- and three-point turns, emerging with small rivers of sweat streaming from their brows. None of which means the Ferrari isn't enjoyable. It *is.* But it's enjoyable like a twenty-minute one-on-one basketball game is enjoyable, and it requires roughly the same concentration.

I found myself double-clutching, matching revs, and moving the shift lever gingerly, as if I were cueing the tone arm on my favorite Sibelius LP. Sure, an Italian supercar deserves that much. But the pindling metal shift rod and its hard plastic knob are clanging throwbacks. Why does Ferrari work so diligently at affecting eccentricity?

However, I also disagree that the Mondial looks old-fashioned. Quite the contrary: I think it's Ferrari's most elegant car. The 328 resembles a Fiero; the Testarossa suffers from hyperglandular gaudiness; the GTO is angular and angry. The Mondial's shape is perfect—the sort of car Candice Bergen might drive, if she were muscular enough to horse it around. In short, the Mondial is the one Ferrari that causes heads to turn in appreciation, rather than shock. But drive it across the country? No, sir, no.—John Phillips III

Mondial interior: Ferrari's most comfortable up front; luggage only in back.

The trick is to time the shifts, and to move the lever with brisk authority. I actually shift by the numbers, unconsciously counting off the discrete moves of the one-two shift, or the four-three shift, against a mental metronome. Jean Lindamood gives the throttle a little blip just as she's sliding the lever into the next gear. Kevin Smith's method is to use only as much clutch travel as is required to disengage. John Phillips III said the shift knob hurt his hand, and we all hooted and jeered and called him a sissy.

Our test car's speedometer was about ten percent fast. I drove along for about 150 miles marveling at its smoothness and silence at ninety miles per hour, until I did a ten-mile check using my watch and the Interstate mileage markers and discovered that I wasn't going ninety miles per hour at all. Other than that, everything on the Mondial worked exactly as it was supposed to during the ten days we had the car.

After my drive from New York, the Mondial was put into our regular test car rotation, and driven daily by all of the Ann Arbor–based writers. We also took turns driving the car on a seventy-eight-mile loop of winding rural roads just west of our home base. Except for Mr. Phillips's sore hand, our notes show remarkable unanimity.

We all found the steering to be extremely heavy at parking speeds and near perfect at highway speeds. The turning circle is a large thirty-nine feet, the steering is three and a quarter turns, lock to lock, and the result is a lot of winding at low speeds. There is also a fair bit of kickback in the steering on bumps. We all found the ride to be excellent, except for a tiresome inability to handle expansion joints and other sharp impacts. Jean Lindamood said, "I wouldn't think twice about knocking off a cross-country trip in a Mondial, but I'd do it on back roads to stay away from those Interstate expansion joints."

The Mondial's brakes proved to be everything one could wish in a car capable of a 140-plus top speed. Modulation was outstanding, especially when the car was being driven hard. We all used the term "squeezing on the brakes" to describe the sensation of brake control that we felt on unfamiliar winding roads. Ferrari's management sniffs contemptuously at the idea of anti-lock brakes, saying that they've already designed a braking system that is exactly right for the car, so why would they want ABS? They are similarly unforthcoming on subjects like four-wheel steering, four-wheel drive, and active suspension.

The Mondial's V-8 engine was admired by one and all. It is one of a very few nonracing engines that are as beautiful to look at as they are to drive. A real shooter would buy his Ferrari with a spare engine that he could keep in his den to admire when he wasn't going anywhere. Unfortunately, Ferrari's very modern foundry only makes parts enough for twenty-four V-8 engines a day, and six or seven of those are shipped to Lancia for installation in the Thema, so a spare might be hard to obtain. The engine is a 90-degree V-8 with four belt-driven overhead camshafts and four valves per cylinder. Ignition system is Marelli-Microplex, and fuel management is Bosch K-Jetronic. The engine produces 260 horsepower at 7000 rpm and 213 pounds-feet of torque at 5500. Maximum allowable revs, 7700. It is an extremely willing engine that pulls strongly right through the rev range and makes a very nice sound in the process. A sound, we might add, that drowns out the stereo at anything above idle.

One area where we failed to agree was front seats. There's no argument about the rears, which are strictly for luggage and grocery bags. But John Phillips III and I were apparently the only members of the staff who liked the Mondial's front

The most beautiful, most musical V-8 engine this side of Cosworth or Ilmor.

seats. Everybody else pronounced them too flat, too lacking in lateral support. After my New York–Detroit dash I was actually quite pleased with them, thinking that they tended to disprove the notion that a driver's seat must look like the one in the Toyota Supra, or a Recaro, in order to be taken seriously.

We are agreed that the driving experience in the Mondial is not a matter of blinding speed, but rather the way Ferrari does it. With a 0-to-60 time somewhere between 7.1 and 7.5 seconds, depending on who's driving, the Mondial will get blown off by several American cars, two or three Porsches, at least one Mercedes, and a couple of BMWs. But none of these feels or sounds or looks like the Mondial. Ferraris are unique, however many seats they have, and that dynamic exclusivity sets them apart from other cars, and is at the same time very difficult to quantify.

Among four-seaters, the Porsche 928S 4 probably comes closest to doing what the Mondial does in the way that the Mondial does it. But we doubt that Porsche 928 prospects are Ferrari Mondial prospects. Similarly, the Mercedes-Benz 560SEL is designed to perform the same mission as the Mondial—with greater comfort, at least equal performance, and an equally heart-stopping price tag—but the respective personalities of those two cars and their buyers are also light-years apart. Until there's another Lamborghini four-seater, the Mondial seems to be the only car in its category.

Should you consider owning one? Of course, provided you genuinely enjoy driving, and you're prepared to go the extra distance to alter your own automotive prejudices to suit the car's idiosyncrasies. Though it's a bit less zoomy, the 3.2 Mondial has the comfort and luggage space that the 328GTS and GTB lack, and it is an eminently more desirable car than the older, larger, and more expensive 412 (not sold in the U.S.). With the 3.2 engine, it now has enough performance to really take advantage of the superior handling and roadholding inherent in its chassis. It may not hold its value, or appreciate, in the same way that some more exciting Ferrari models have in the recent past, but you're not going to lose a lot of money on any Ferrari. All that remains in question is your relationship with your bank. STOP

FERRARI MONDIAL 3.2

GENERAL:
Mid-engine, rear-wheel-drive coupe
2 + 2-passenger, 2-door steel body
Base price $71,700

MAJOR EQUIPMENT:
Air conditioning standard
Sunroof standard
AM/FM/cassette not available
Leather interior standard
Cruise control not available

ENGINE:
32-valve DOHC V-8, aluminum block and heads
Bore x stroke 3.27 x 2.90 in (83.0 x 73.6mm)
Displacement 194 cu in (3185cc)
Compression ratio 9.2:1
Fuel system Bosch K-Jetronic mechanical injection
Power SAE net 260 bhp @ 7000 rpm
Torque SAE net 213 lb-ft @ 5500 rpm
Redline 7700 rpm

DRIVETRAIN:
5-speed manual transmission
Gear ratios (I) 3.42 (II) 2.35 (III) 1.69 (IV) 1.24 (V) 0.92
Final-drive ratio 4.06:1

MEASUREMENTS:
Wheelbase 104.3 in
Track front/rear 59.8/59.4 in
Length 178.5 in
Width 70.7 in
Height 48.6 in
Curb weight 3400 lb
Fuel capacity 18.5 gal

SUSPENSION:
Independent front, with upper and lower A-arms, coil springs, anti-roll bar
Independent rear, with upper and lower A-arms, coil springs, anti-roll bar

STEERING:
Rack-and-pinion

BRAKES:
11.1-in vented discs front
11.0-in vented discs rear

WHEELS and TIRES:
16 x 7.0-in front, 16 x 8.0-in rear cast magnesium wheels
205/55VR-16 front, 225/55VR-16 rear
Goodyear Eagle VR55 tires

PERFORMANCE (manufacturer's data):
0–60 mph in 7.4 sec
Standing ¼-mile in 15.0 sec
Top speed 149 mph
EPA city driving 13 mpg

MAINTENANCE:
Headlamp unit $50.38
Front quarter-panel $551.18
Brake pads front wheels $98.92
Air filter $33.25
Oil filter $16.47
Recommended oil change interval 7500 miles

	POOR	FAIR	GOOD	EXCELLENT
ENGINE				
power				•
response				•
smoothness				•
DRIVETRAIN				
shift action		•		
power delivery				•
STEERING				
effort	•			
response				•
feel			•	
RIDE				
general comfort			•	
roll control				•
pitch control				•
HANDLING				
directional stability				•
predictability				•
maneuverability			•	
BRAKES				
response				•
modulation				•
effectiveness				•
GENERAL				
ergonomics			•	
instrumentation			•	
roominess			•	
seating comfort			•	
fit and finish				•
storage space				•
OVERALL				
dollar value			•	
fun to drive				•

FERRARI FABRICATION

Computer-guided automation coexists with hammers and sweat in the making of a legend.

PHOTOGRAPHY BY DOUGIE FIRTH

In the manufacture of Ferraris, time and progress are put to use where they can help, ignored where they cannot. And always, the sense of community remains. Enzo Ferrari once said: "Here it is always like a family. Here worked the father, often the mother, the son, the daughter-in-law, the uncle. They feel as if they are at home and that the factory has remained a part of the family and that its fate is their fate."

Ferrari has always been willing to farm out body welding and to buy minor components anywhere, but the heart of the car—the engine—is purely an in-house creation. Automated machining operation prepares raw castings for final assembly.

Only 2500 cars roll out the Maranello gates in a year—not so many that each one can't have minor flaws tended to with a well-placed hammer.

The aluminum DOHC, 32-valve V-8 takes shape at a couple of spotless, neat work stations. In its 3.2-liter, 260-bhp specification, it is installed in the Mondial, the 328GTB, and the 328GTS. A few 215-bhp 3.0s are built for Lancia's Thema sedan.

Don't tell OSHA: Welding practices in Ferrari body construction, including the use of much lead, would never stand up to American work rules and regulations. But, hey, it's Italy.

Compared with the thoroughly modern foundry and the reasonably up-to-date assembly lines, the Scaglietti body works seems crude and old-fashioned. Steel Mondial bodies are fabricated by a bunch of guys with hammers and tin snips.

Ferrari 3.2 MONDIAL

Eight cylinders, 32 valves, 270bhp, a mid-mounted engine and four seats promise performance with relative practicality in Ferrari's Mondial. The promise is kept

PHOTOGRAPHY BY PETER BURN

Ferrari 3.2 MONDIAL

AT A GLANCE

FOR

Exciting performance from charismatic V8; safe, enjoyable handling; good ride; roomy and practical for a mid-engined car; painstaking build quality

AGAINST

Poor fuel economy; indifferent ventilation; noisy if you're not in the mood; rather expensive

PRICE	£48,102
MAX SPEED	148.5mph
0-60MPH	6.3sec
MPG OVERALL	15.9
TEST DISTANCE	582 miles

Two years ago we tested a Ferrari 328 GTB. We drooled over its sensuous looks; we eulogised its electrifying performance; we thrilled to the sabre-sharp handling; we soaked up its overflowing charisma. We loved that car, noisy, impractical and expensive though it was. But what if you need, sometimes, to carry more than just one passenger? Are you to be denied the pleasures of Ferrari's 3185cc 32-valve V8 with the searing scream that comes from the firing order dictated by a flat-plane crankshaft? Does it mean front-wheel drive saloon luxury, a detuned version of that V8 (with a 90deg crank and a more conventional burble) – in short, a Lancia Thema 8.32 for you?

Were it so, it would surely be no hardship. But fear not; bridging the gulf is Ferrari's own Mondial, whose mechanical make-up is practically a carbon copy of the 328's. It offers, possibly uniquely, four seats within a mid-engined configuration. Clothing a tubular chassis, the Mondial's crisp, heavily louvred, Pininfarina styling is unmistakably Ferrari (and disguises the long wheelbase well); but as a car to live with day to day it promises to be as practical as a Porsche.

The Mondial has been around since 1980, the year it replaced the Bertone-styled 308 GT4. Since then, the Mondial and the 308/328 GTB and GTS have jointly benefited from successive rounds of improvements, with *quattrovalvole* cylinder heads in 1982 and a hike to the present engine capacity (from 2927cc) in 1986.

Not surprisingly, the Mondial is 11in longer than the 328, and nearly three inches wider too; it fits its extra seats within a 104.3in wheelbase compared with the 328's 92.5in. It is also more expensive, at a heady £48,102 compared with £44,199 for the 328 GTB. If you would prefer your Mondial in cabriolet guise, the outlay rises to £52,300. A Porsche 911, even with Sport Equipment, begins to look almost a bargain at £38,903 – and that car, also an occasional four-seater, is the Ferrari's greatest adversary.

PERFORMANCE AND ECONOMY

The Mondial weighs 2.2cwt more than the 328 GTB, and has a greater frontal area. It's not surprising, then, that it is slightly less rapid. Nevertheless, the V8's 270bhp is sufficient to propel the Mondial's 28.3cwt to a Millbrook maximum of 148.5mph, exactly 10mph down on the speed we recorded for the GTB and fractionally past the 7000rpm power peak in fifth gear. Tyre scrub absorbs a good 5mph when a car laps the Millbrook bowl at this sort of speed so the Mondial is still, potentially, a true 150mph car. It will get there, furthermore, with an animal ferocity which is uniquely Ferrari. An awkward dog-leg gearlever movement adds valuable milliseconds to the up-change into second gear, but there's not too much wrong with a 6.3sec 0-60mph time (0.8sec slower than the GTB's). Nor, indeed, with the 15.8sec needed to hit 100mph, or the paltry 5.7sec taken to touch the legal limit from 30mph. *That* is the measure of how quick the Mondial can be on the road.

You can take the engine up to the 7750rpm red line with ease, with only the moderating influence of the rev limiter to stop it heading beyond 7800rpm. Beginning as a humdrum hum overlaid with a four-cylinder Fiat-like whine at low revs, the excitement of the engine's mechanical song mounts as the revs rise. At 5000rpm the note has hardened and the power floods in; at 6000rpm it's a neck-tingling howl, beyond 7000rpm it screams encouragement. The whirrs and whines and the rustle of 32 reciprocating valves are all drowned in an aural explosion of potency. This is a wonderful engine.

And smooth with it. Firing intervals 180deg apart mean, in theory, that the engine shouldn't be as smooth as a V8 usually is, but the practice is different. The V8 spins like a turbine, with a throttle response as crisp as the best. It's flexible, too; the meek voice at low revs disguises a pick-up so muscular you can regularly hold a gear higher than you'd have thought possible. There's a meaty shove from less than 1000rpm, and no trace of temperament; in fourth gear, all the increments from 30-50mph through to 70-90mph are disposed of in around five seconds or less. It is here, more than in outright acceleration and speed, that this latest Mondial scores over the previous, smaller-engined version.

That the Mondial is a short-geared car (20.9mph/1000rpm in fifth, 15.5mph/1000rpm in fourth) helps its tractability; while the 223lb ft torque peak doesn't appear until 5500rpm, the torque spread reaches far down and, with such a torrent of power at very high engine speeds, the gearing is exactly right for the engine.

What it needs is a gear-change to match its responsiveness. The slender chrome lever clacks through Ferrari's familiar open metal gate smoothly enough once the gearbox has warmed up (a double-declutch helps when the oil is cold), but it needs a firm hand to effect a fast change and a sympathetic ear to effect a clean one. Rapid motoring calls for fast gear-changes because the ratios are

The Works

Heart of the Mondial is the all-aluminium 90deg V8, with 16 valves in each of its twin-cam cylinder heads. Its oversquare, 83 × 73.6mm bore/stroke dimensions — both increased over the first *quattrovalvole* — give it a swept volume of 3185cc. Ferrari remain faithful to Bosch mechanical fuel injectio (K-Jetronic) but use a fully mapped Marelli Microplex ignition system.

The use of a flat-plane crankshaft is very unusual in a production V8, though it is common in racing engines. It allows simpler inlet and exhaust plumbing and hence greater power outputs, at the theoretical expense of some smoothness. To illustrate the point, the engine used in the Lancia Thema 8.32, basically similar apart from a crankshaft with 90deg firing interval, gives 215bhp and 209lb ft of torque compared with 270 and 223 respectively for the Ferrari.

Also unusual is the positioning of the transaxle behind the transverse engine, with drive taken to it via an idler wheel.

closely, if unevenly, spaced; first and second are rather close, and there's quite a gap between fourth and fifth. The clutch however is smooth, progressive and has good feel.

Short gearing and good fuel economy seldom mix, but even so 15.9mpg overall is a disappointment. A touring 21.0mpg offers a range of around 400 miles from one 19.1-gallon, tankful of four star.

Handling and Ride

Naturally the Mondial follows regular Ferrari suspension practice, with coil-sprung double wishbones at each corner, anti-roll bars at each end and rack and pinion steering. With a slightly tail-heavy static weight distribution of 46 per cent front, 54 per cent rear, the Mondial now uses larger tyres for the driven wheels: 240/55 VR 390 compared with 220/55 VR 390 for the front, with wider rims to suit.

The end result is a car which is probably the most forgiving of all mid-engined cars. The long wheelbase helps here, taming the responses away from the snap breakaway of some mid-engined machines (GTB included) and giving a driver-friendly balance the pendulous 911 lacks.

First impressions are not promising. The Ferrari is taut, but its steering feels low-geared (it is) and there's a lot of understeer if you enter a bend under gentle power. Even easing the throttle fails to quell it completely. Apply more power, though, or turn-in on a feathered throttle, and instead of the expected build-up of understeer the balance changes to neutrality, even at moderate speeds. Further power will push the tail out into gradual oversteer, easily modulated either with the throttle or by winding off a little steering lock.

Should the tail move out too far, or you are forced to lift off, you'll need to apply a measured amount of opposite lock before doing anything more with your right foot. The Ferrari will then flow obediently back on to its line with no drama. It's a safe, forgiving chassis which inspires great confidence; you can hold the Mondial in an oversteering slide in a way only the very brave would attempt in a 911, mindful that left to its own devices the Ferrari will naturally revert to stabilising understeer.

The steering, ideally weight-

ed, precise and always informative of the road beneath the front tyres to the extent of kicking back slightly, becomes beautifully responsive when the Mondial is set up to quell the understeer. Yet it is this initial, artificial understeer which smacks of compromise, of making the mid-engined, highly potent Mondial safe in the hands of road drivers who may not be highly skilled. There is too much, and the contrast between it and the neutrality when going harder is too great. To get the best out of the Mondial, you have to work harder than you would in, say, a Lotus Excel or a Porsche 944 Turbo. Do that, though, and the rewards are great.

The Mondial's tautness is reflected in a firmly damped ride and low roll angles. High-speed body control is superb, and around town the Ferrari doesn't fidget. Even rear seat passengers approve. The brakes, too, are fine with a short travel, good feel and a progressive action, ABS, standard for the UK market, overcomes the criticism of the previous Mondial that it could lock its front wheels in the wet.

INTERIOR

First impression inside the Mondial is of airiness. The pillars are slim, the scuttle is low, the view aft is unobstructed. There can be few mid-engined cars to compare with the Mondial here. The cabin is wide too, and with surprisingly good headroom the rear seats are habitable by compactly built adults for short distances provided they don't mind splaying their knees. They're fine for children.

The front seats, leather-trimmed like most of the interior, are firm but their slippery surface doesn't aid lateral support. The head restraints, moreover, are mounted too low. Seat adjustment, for runner position and backrest rake only, is achieved manually; the rival Porsche's adjustments are electric and include cushion height/tilt.

Pedals offset to the left make the driving position strange initially, but it proves comfortable in practice. Heel-to-toe gearchanges are straightforward and there's a rest for the left foot. Fine tuning is achieved with a steering wheel adjustable for reach and height. Invariably, though, testers found that the tops of the instruments were obscured when the three-spoke leather-rimmed wheel was positioned to their preference. It seems that the instrument binnacle, well stocked with clear, bold red-on-black dials and an unusual 'detached' odometer/trip meter, would be better mounted a little lower.

Three Fiat-sourced stalks sprout from the steering column, and solenoid switches next to the wildly optimistic speedometer open the bonnet (full of spare wheel and reservoirs), engine lid (the engine is surprisingly accessible) and glassfibre boot lid (with usefully sized boot). Further switches between the front seats open the glovebox (only when the ignition is on!), the windows and control the air conditioning. This last item will not allow a cool face and warm feet at the same time.

Low scuttle and slim pillars help give the Mondial a light and airy look. Steering wheel is adjustable for reach and height

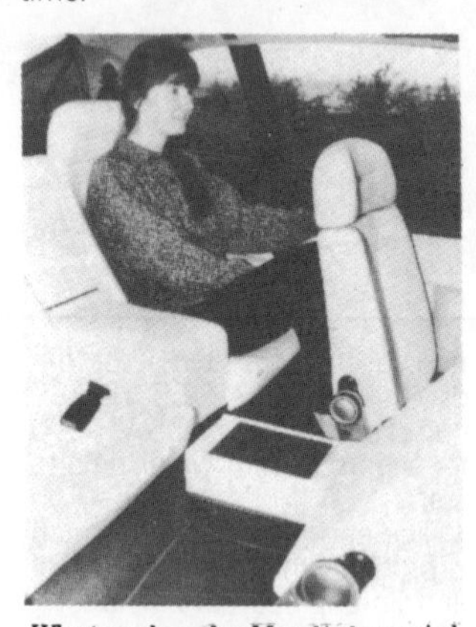

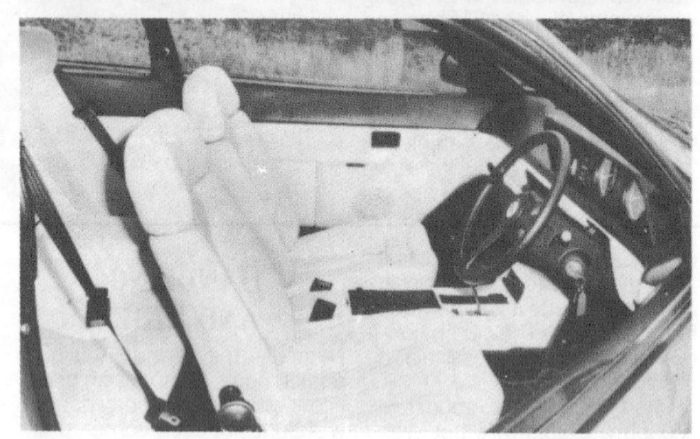

What makes the Mondial special . . . a mid-engined configuration with enough passenger space for adults in the rear compartment. Cream leather upholstery is beautifully presented

The test car's gearbox developed a loud whine during our maximum speed run, exactly as happened with our test GTB. Oil surge on the banking appears to be the cause. Consequently we couldn't record the cabin noise levels. However, it doesn't take a noise meter to tell you that the Mondial's cabin is a pretty noisy place. The engine's whirrings are ever-present, though ultimately drowned by the more stirring sounds we've described. Wind noise, road roar and suspension thump can't compete; coasting in neutral shows that all are well muted anyway.

FINISH

There's no doubt about it: the Ferrari is beautifully made. Our test car's red paint was practically flawless, and such little unevenness in the cream leather upholstery as there was actually added to the bespoke look. The only black mark was that the lining to the first-aid kit cover flap, between the rear seats, had become unglued. The overall feel of the interior blends Ferrari sportiveness with luxury very effectively; it's opulent, but understated.

Outside, body-colour bumpers make for a tidier appearance than in earlier Mondials, while those torpedo-shaped etched chrome door handles are, of course, a typically Pininfarina touch.

EQUIPMENT

Central locking (from either door), a Blaupunkt stereo, electrically adjustable mirrors, air conditioning and tinted glass are standard, plus of course the ABS braking and the leather interior already mentioned. It's enough for a car like this; trip computers and cruise controls have no place in a Ferrari. You buy a car like this because it wears the Prancing Horse, not because of the toys thrown in. Extra leather and an electric sunroof are items for the options list.

COSTS

Who can tell? Maintaining a Ferrari will never be cheap, and regular servicing is vital. Just remember that 32 valves add up to a lot of tappet clearances to adjust.

Nor is the Mondial notably economical on fuel; arguably, if you can afford to buy it, you can afford to fuel it. Depreciation is hard to predict, as shown by the way early Dinos and late GT4s have shot up in value alike. Look after your Mondial, and it will repay you handsomely in years to come.

The Verdict

Of course the Mondial, like all supercars, is impossible to justify objectively, but not everyone buys a car on that basis. Yes, a Ford Sierra Cosworth does most of what a Mondial can do at under half the cost, but that misses the point completely. With a Ferrari, you're buying the sound, the look, the reputation, the name, a piece of fine engineering, the sheer driving excitement the car offers.

So our ratings chart has to be looked at in the context of the Ferrari being an inherently expensive, hand-built, mid-engined supercar. Given what it is, the accommodation is brilliant. Nevertheless, people are most likely to buy the Mondial because of the extra space.

Is the Mondial better than a Porsche 911? In some ways, yes; it's more forgiving to drive both slowly and fast, and it has more luggage room. A 911 is quicker, cheaper and will probably cost less to maintain. Both are entirely practical for day-to-day use, and in this the Mondial is rare among Ferraris.

That it can even come down to a rational choice proves one point, though: in the Mondial, Ferrari have made a car for the real world. M

Rating

Performance	■ ■ ■ ■ ■ ■ ■ □ □ □
Economy	■ ■ ■ ■ □ □ □ □ □ □
Transmission	■ ■ ■ ■ ■ □ □ □ □ □
Handling	■ ■ ■ ■ ■ ■ ■ ■ □ □
Brakes	■ ■ ■ ■ ■ ■ ■ ■ □ □
Ride comfort	■ ■ ■ ■ ■ ■ ■ ■ □ □
Accommodation	■ ■ ■ ■ ■ ■ ■ ■ □ □
Boot/storage	■ ■ ■ ■ ■ ■ □ □ □ □
At the wheel	■ ■ ■ ■ ■ ■ ■ □ □ □
Visibility	■ ■ ■ ■ ■ ■ ■ ■ ■ □
Instruments	■ ■ ■ ■ ■ ■ □ □ □ □
Heating	■ ■ ■ ■ ■ ■ □ □ □ □
Ventilation	■ ■ ■ ■ ■ □ □ □ □ □
Noise	■ ■ ■ ■ ■ ■ □ □ □ □
Finish	■ ■ ■ ■ ■ ■ ■ ■ ■ □
Equipment	■ ■ ■ ■ ■ ■ □ □ □ □

Motor's Rating

■ ■ ■ ■ ■ ■ ■ ■ □ □

Value Rating

■ ■ ■ ■ ■ □ □ □ □ □

Standard Equipment

Seat back map pockets	
Map reading light	■
Boot light	■
Central door locking	■
Remote boot/hatch release	■
Remote fuel flap release	■
Remote control door locking	
Electric mirror adjustment	■
Heated mirrors	
Intermittent wipe (variable)	■
Programmed wash/wipe	
Headlamp wash/wipe	
Internal headlamp levelling	
Driving lamps	
Fog lamps	■
Radio/cassette player	■
Electric aerial	■
Electric sunroof	□
Illuminated vanity mirror	
Seat height adjustment (driver)	
Seat tilt adjustment (driver)	
Adjustable upper seatbelt mounting	
Adjustable steering column	■
Rear seat head restraints	
Rear centre armrest	■
Anti-lock braking system	■
Rear compartment heating	
Tinted glass	■
Air conditioning	■
Power assisted steering	
Self levelling suspension	
Cruise control	
Limited slip differential	■
Leather trim	■

□ Options

Costs and Service

Insurance group	9
Major service interval, miles	12,500
Intermediate service, miles	6250
Oil change, miles	—
Set brake pads (front) £	51.44
Complete clutch £	244.89
Complete exhaust £	634.11
Front wing panel £	317.30
Oil filter £	9.82
Starter motor £	393.61
Windscreen £	411.52
Tyre £*	156.58†
Total service time, hrs up to 50,000 miles	N/A
Time to change clutch, hrs	3.5

*Discounted price †rear; front £123.73

Make and Model

Ferrari 3.2 Mondial

MADE BY: **Ferrari S.E.F.A.C.**

AT: **Maranello, near Modena, Italy**

SOLD IN THE UK BY: **Maranello Sales Ltd, Egham Bypass (A30), Egham, Surrey TW20 0AX**

TEL: **0784 36431**

NUMBER OF DEALERS: **16**

Price

TOTAL PRICE: **£48,102**

EXTRAS FITTED TO TEST CAR: **Electric sunroof £1127, Leather dashboard £810**

OTHER OPTIONS: **Leather headlining and rear window surround £573**

PRICE AS TESTED: **£50,039**

Performance

WEATHER CONDITIONS

Wind	4mph
Temperature	58degF/14degC
Barometer	29.9in Hg/ 1013mbar
Surface	Dry tarmacadam

MAXIMUM SPEEDS

	mph	kph
Banked circuit (5th gear)	148.5	238.9
Terminal speeds:		
at ¼ mile	97.4	156.7
at kilometre	122.5	197.1
Speeds in gears (at 7750rpm):		
1st	43	70
2nd	63	101
3rd	88	142
4th	120	193

ACCELERATION FROM REST

mph	sec	kph	sec
0-30	2.5	0-40	2.0
0-40	3.4	0-60	3.2
0-50	5.0	0-80	5.0
0-60	6.3	0-100	6.8
0-70	8.2	0-120	9.0
0-80	10.0	0-140	11.8
0-90	11.9	0-160	15.6
0-100	15.8	0-180	20.3
0-110	19.2	0-200	27.8
0-120	23.9		
Stand'g ¼	14.8	Stand'g km	26.7

ACCELERATION IN TOP

mph	sec	kph	sec
20-40	8.0	40-60	5.0
30-50	7.6	60-80	4.7
40-60	7.1	80-100	4.5
50-70	7.2	100-120	4.6
60-80	7.7	120-140	5.2
70-90	8.3	140-160	5.9
80-100	9.3		

ACCELERATION IN 4TH

mph	sec	kph	sec
20-40	5.4	40-60	3.3
30-50	5.0	60-80	3.0
40-60	4.8	80-100	3.0
50-70	4.8	100-120	3.0
60-80	4.9	120-140	3.3
70-90	5.1	140-160	3.3
80-100	5.3	160-180	4.6
90-110	6.3		
100-120	8.7		

FUEL CONSUMPTION

Overall	15.9mpg
Touring*	21.0mpg
Govt tests	13.6mpg (urban) 31.4mpg (56mph) 25.2mpg (75mph)
Fuel grade	97 octane 4 star rating
Tank capacity	19.1gal 87 litres
Max range*	401 miles
Test distance	582 miles

*Based on official fuel economy figures – 50 per cent of urban cycle, plus 25 per cent of each of 56/75mph consumptions.

STEERING

Turning circle	41ft 7in 12.7m
Lock to lock	3.4 turns

NOISE Not measured (see text)

SPEEDOMETER (MPH)

True mph	30	40	50	60	70	80	90	100
Speedo	36	46	56	67	79	90	101	111

Distance recorder: 1.6 per cent fast

WEIGHT

	cwt	kg
Unladen weight*	28.3	1438
Weight as tested	32.5	1653

*No fuel

Performance tests carried out at 2054 miles by *Motor's* staff at the Millbrook proving ground, near Ampthill.

Specification

ENGINE

Cylinders	V8
Capacity	3185cc
Bore/stroke	83/73.6mm
Max power	270bhp (199kW) at 7000rpm (DIN)
Max torque	223lb ft (304Nm) at 5500rpm (DIN)
Block	Aluminium alloy
Heads	Aluminium alloy
Valve gear	Dohc per bank, toothed belt drive, bucket tappets, four valves per cyl
Compression	9.8:1
Fuel system	Bosch K-Jetronic fuel injection
Ignition	Marelli Microplex
Main bearings	Five

TRANSMISSION

Layout	Tranverse mid-engine, rear-wheel drive
Type	Five-speed, manual, limited-slip differential
Internal ratios and mph/1000rpm	
Top	0.92/20.9
4th	1.24/15.5
3rd	1.69/11.4
2nd	2.35/8.1
1st	3.42/5.6
Rev	3.25
Final drive	3.82:1

AERODYNAMICS

Cd N/A

SUSPENSION

Front	Double wishbones, coil springs, anti-roll bar
Rear	Double wishbones, coil springs, anti-roll bar

STEERING

Type	Rack and pinion
Assistance	No

BRAKES

Front	Vent'd discs 11.4in dia
Rear	Vent'd discs 11.8in dia
Servo	Hydraulic pump
Circuit	Split front/rear, ABS
Handbrake	On rear wheels

WHEELS/TYRES

Type	Alloy, 165 TR × 390mm dia front, 180 TR × 390mm dia rear
Tyres	Michelin TRX, 220/55 VR 390 front, 240/55 VR 390 rear
Pressures F/R (all conditions)	34/34psi 2.4/2.4bar

GUARANTEE

Duration	12 months, unlimited mileage
Rust warranty	Coupon plus annual checks

MAINTENANCE

Major service	12,500 miles
Intermediate	6250 miles

THE RIVALS

Others include the Aston Martin V8 (£69,500), Audi Quattro (£30,199), Lotus Esprit Turbo (£29,950) and Renault GTA Turbo (£26,990)

FERRARI 3.2 MONDIAL £48,102

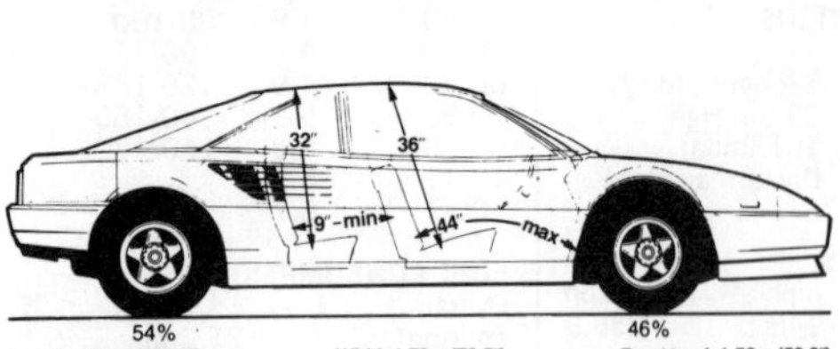

54% 46%

Length 4·58m (180·3") Width 1·79m (70·5") Front track 1·52m (59·8")
Wheelbase 2·65m (104·3") Height 1·26m (49·5") Rear track 1·49m (58·8")

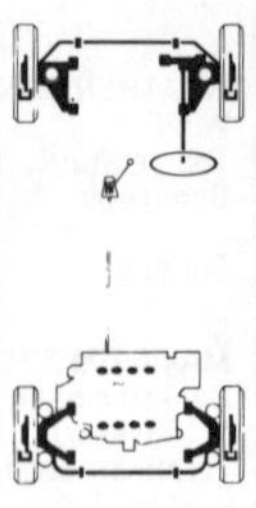

Capacity, cc	3185
Power bhp/rpm	270/7000
Torque lb ft/rpm	223/5500
Max speed, mph	148.5
0-60 mph, sec	6.3
30-50 mph in 4th, sec	5.0
30-70 mph through gears	5.7
mph/1000 rpm	20.9
Overall mpg	15.9
Touring mpg	21.0
Weight cwt	28.3
Boot capacity, ft^3	N/A
Drag coefficient, Cd	N/A

Ferrari's token four-seater shares the super-smooth wailing 32-valve V8 that powers the 328 GTB, and while greater weight blunts performance the Mondial is still a very quick car. Economy is poor, but handling is well balanced for hard driving, the brakes are excellent and the ride supple. The most practical of mid-engined supercars, very noisy but beautifully made. A Ferrari you can live with day to day.

BMW M635 CSi £45,780

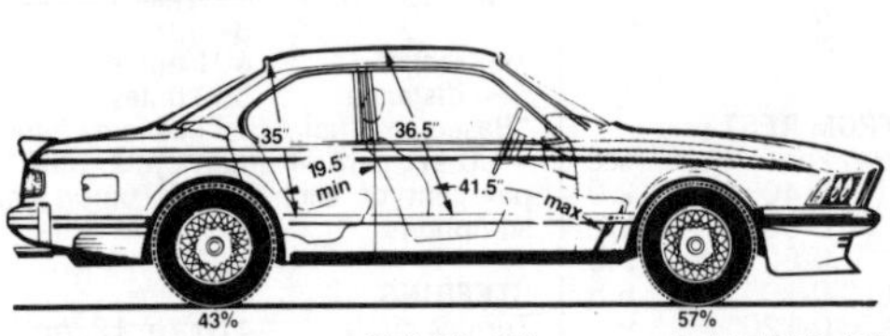

43% 57%

Length 4·75m (187") Width 1·73m (68") Front track 1·42m (56")
Wheelbase 2·62m (103") Height 1·37m (54") Rear track 1·46m (57·5")

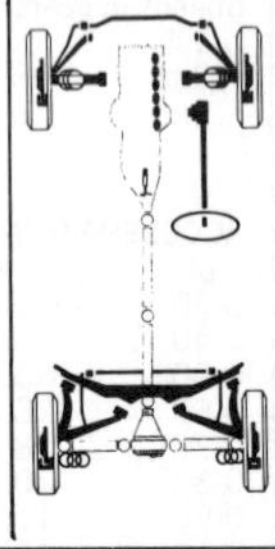

Capacity, cc	3453
Power bhp/rpm	286/6500
Torque lb ft/rpm	251/4500
Max speed, mph	149.7
0-60 mph, sec	6.3
30-50 mph in 4th, sec	6.7
30-70 mph through gears	5.4
mph/1000 rpm	23.8
Overall mpg	16.9
Touring mpg	24.0
Weight cwt	29.6
Boot capacity, ft^3	12.3
Drag coefficient, Cd	n/a

Fastest BMW available in the UK takes the decade-old 6-Series firmly into supercar territory and more than holds its own. Fabulous 24-valve six delivers lusty performance with good refinement but is very thirsty. Handling is both enjoyable and forgiving, ride firm but well-controlled. Interior is too ordinary for the price but build and finish are first class. Generally well equipped but air conditioning is extra.

JAGUAR XJ-S 3.6 £24,700

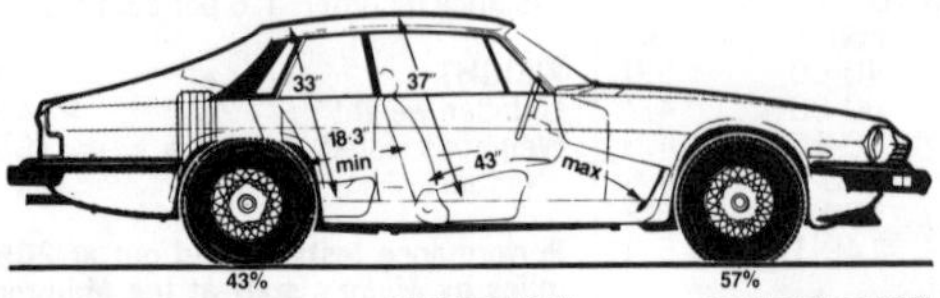

43% 57%

Length 4·87m (191·8") Width 1·79m (70·5") Front track 1·49m (58·5")
Wheelbase 2·59m (102") Height 1·26m (49·8") Rear track 1·47m (58")

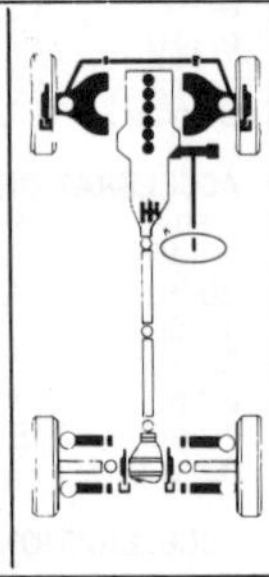

Capacity, cc	3590
Power bhp/rpm	225/5300
Torque lb ft/rpm	240/4000
Max speed, mph	136.8
0-60 mph, sec	7.2
30-50 mph in 4th, sec	6.9
30-70 mph through gears	7.0
mph/1000 rpm	28.9
Overall mpg	18.9
Touring mpg	24.2
Weight cwt	32.5
Boot capacity, ft^3	10.9
Drag coefficient, Cd	0.40

Jaguar's smaller-engined six-cylinder XJ-S excels in most areas. Its 3.6-litre AJ6 engine is now superb, having benefited from development of the unit for its application in the new XJ6. The rest is good news, too: a positive gearchange, a beautifully balanced chassis with good grip and a superb ride, fine finish and appointments. New Sport suspension pack gives tauter responses. Still a desirable, if indulgent, 2+2.

LAMBORGHINI JALPA £43,656

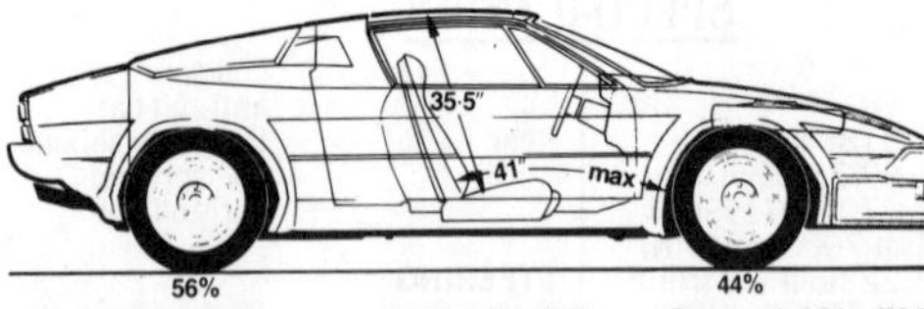

56% 44%

Length 4·22m (166") Width 1·65m (65") Front track 1·54m (60·5")
Wheelbase 2·45m (96·5") Height 1·12m (44") Rear track 1·48m (58·5")

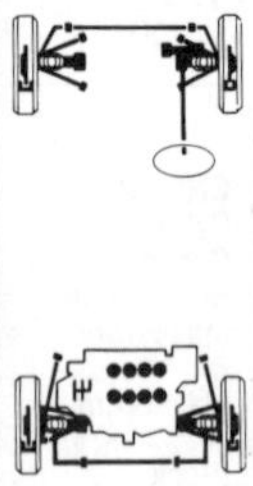

Capacity, cc	3485
Power bhp/rpm	250/7000
Torque lb ft/rpm	235/3250
Max speed, mph	147.6
0-60 mph, sec	5.8
30-50 mph in 4th, sec	4.3
30-70 mph through gears	5.5
mph/1000 rpm	20.4
Overall mpg	15.6
Touring mpg	17.6
Weight cwt	26.6
Boot capacity, ft^3	n/a
Drag coefficient, Cd	n/a

'Baby' of the two-car Lamborghini range, the targa-top Jaina's ancestry runs back to the mid-engined Uracco of the early '70s. Magnificently vocal quad-cam V8 delivers fine performance with massive mid-range punch, though economy is mediocre by today's standards. Very safe and ultimately forgiving handling married to reasonable ride. Fabulous brakes. He-man gearchange and poor visibility not so appealing.

PORSCHE 911 SE £38,903

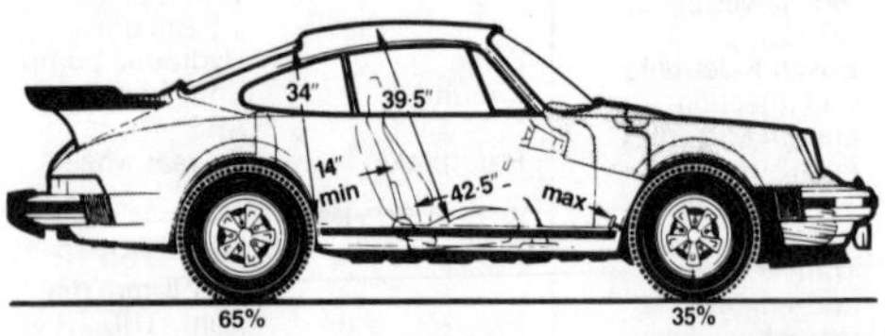

65% 35%

Length 4·29m (169") Width 1·77m (69·8") Front track 1·50m (59")
Wheelbase 2·27m (89·5") Height 1·32m (52") Rear track 1·43m (56.3")

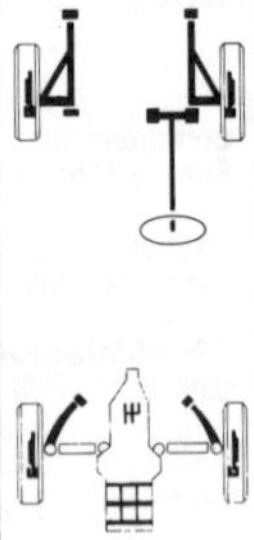

Capacity, cc	3164
Power bhp/rpm	231/5900
Torque lb ft/rpm	209/4800
Max speed, mph	151.1
0-60 mph, sec	5.3
30-50 mph in 4th, sec	5.6
30-70 mph through gears	5.3
mph/1000 rpm	24.3
Overall mpg	21.1
Touring mpg	28.6
Weight cwt	23.0
Boot capacity, ft^3	9.9
Drag coefficient, Cd	0.38

Little changed for the past couple of years but still at the top of the junior supercar acceleration league table, the 911 Carrera is also remarkably economical for its stunning performance. Still a great driving machine, with rewarding handing (though tricky on the limit), potent brakes, superb ratios, good driving position and turbine-smooth engine. Gearchange improved, but remaining flaws include hard ride and poor heating/ventilation.

PORSCHE 928 S4 £54,826

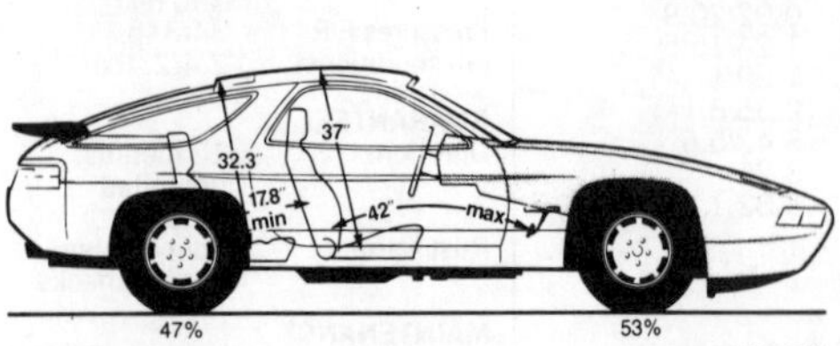

47% 53%

Length 4·44m (175") Width 1·84m (72·3") Front track 1·55m (61")
Wheelbase 2·50m (98·5") Height 1·31m (51·8") Rear track 1·53m (60·3")

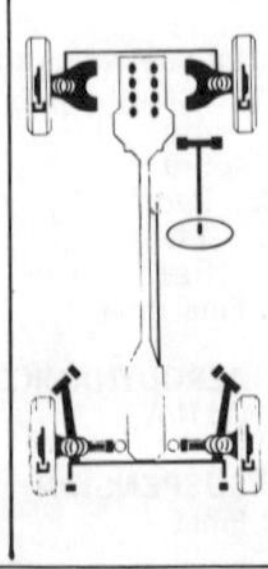

Capacity, cc	4957
Power bhp/rpm	320/6000
Torque lb ft/rpm	317/3000
Max speed, mph	158.7
0-60 mph, sec	6.4
30-50 mph in kickdown, sec	2.2
30-70 mph through gears	5.4
mph/1000 rpm	28.0
Overall mpg	17.4
Touring mpg	22.6
Weight cwt	30.9
Boot capacity, ft^3	7.4
Drag coefficient, Cd	0.34 to 0.352

Latest version of the flagship front-engined Porsche has 32 valves and a full five litres increased power and a more slippery body make for even greater performance combined with surprising economy, though standard-fit automatic transmission blunts the sporting edge. Handling is sharp and fluid but body control is imperfect and low speed ride is harsh. Interior is opulently functional and the S4 is beautifully made, but pricey.

Enzo's new hit

FERRARI'S 348 TWO-seater will be launched in autumn 1989 to replace the current 328 series.

The new car has been photographed testing at Fiorano, still running with light disguise panels at the side and rear. The picture shows that Pininfarina has concentrated on giving the 348 a stronger family resemblance to the Testarossa and Mondial, moving to a lower, flatter line and away from the 328's distinctive curves.

The new car has a longer, lower bonnet and Testarossa-style side strakes. The rear end treatment remains disguised, but the simpler line will be continued with no noticeable curve over the wheelarch.

Although the 348 is slightly larger, sources close to Ferrari says it will be considerably more aerodynamic and probably weigh around the same. There have been few criticisms of the 328's looks, but it weighs in at a comparatively heavy 2975lbs (that is still 80lb lighter than the Lotus Esprit turbo) and is not noted for its wind-cheating qualities.

Ferrari's managing director Giovanni Razelli says that the new car will not be making use of lightweight composites and that it will remain rear-wheel driven. There are no plans to introduce a four-wheel-drive system for the 348.

The biggest mechanical change is the engine. The 32-valve, 3.4-litre, fuel-injected V8 has been developed from the 270bhp 3.2-litre unit, using a slightly enlarged bore and stroke.

The 3.4-litre engine is reputed to develop around 300bhp, but Italian sources say that Ferrari is concentrating on improving low-down torque and making it more compatible with US emissions regulations.

Other mechanical details are typical and traditional Ferrari — double wishbone front and rear suspension and a tubular steel frame.

Razelli says the 348 will not be available until around September 1989 and it will probably make its public debut at next year's Frankfurt show. It will be more expensive than the 328 — early indications suggest a price tag of £50,000. The current 328GTB costs £44,196 and the GTS £45,598.

At Ferrari's annual general meeting on 8 June, it was revealed that the company made a profit of £6.36 million during 1987. Turnover was up by 15per cent to £155 million and the company is spending nearly 10per cent of turnover on research and development.

The board meeting turned out to be less stormy than predicted, following Enzo Ferrari's row with his son, Piero Lardi Ferrari.

It was confirmed that Piero Lardi Ferrari has left the racing team and taken up the position of executive vice president of the production side — Ferrari Sefac SpA.

There have been strong rumours, still unconfirmed, that Enzo Ferrari, now 90 and in failing health, is ready to sell his 40 per cent stake in the company to Fiat.

Fiat already holds 50per cent, with the remaining 10per cent owned by Piero Lardi Ferrari. Under an agreement made in 1969, when Fiat first took a share in Ferrari, Fiat has first refusal on Enzo's shares — and the word from Italy is that Fiat will certainly take up that option.

Gathering of the greats

OVER 30 MILLION pounds worth of Ferrari — 150 cars in total — were tested at the Castle Combe circuit in Somerset last week.

The event was organised by Modena Engineering and attracted some of the most sought-after Ferraris, many of which changed hands during the day.

The most expensive were a brace of 250 GTOs, each worth £1.5 million. They put the new GTO, the 288, in the shade. There were four 288s with a collective value of £1 million.

Others included a Daytona Spider, valued at £500,000 and 10 512 Testarossas — each worth £100,000.

Modena event: visited by 150 cars

FERRARIS

Lancia's Ferrari-engined Thema, just on sale in Britain, battles a car using a similar engine,

FOR FOUR

Maranello's under-rated Mondial, a machine also promising room for four/ Richard Bremner

WHAT'S THIS? LANCIA Thema 8.32 versus Ferrari Mondial? Surely that's like comparing a Renault GTA with a Renault 25 V6 because they share the same engine? But that's not all that binds this pair. Both use basically the same V8 Ferrari engine, yes, and both are from different arms of Fiat's octopus-like empire. But the vital thing that plunges these machines into the same market, even if they arrived from different directions, is that both are four-seater sports cars.

The Mondial is Ferrari's only serious shot at a

sporting four-seater despite regular rumours, and equally persistent denials, that a proper four-door, four-seater is lurking in the shadows at Maranello. True, Ferrari makes that rakish heavy goods vehicle, the 412, also a *quattro posti* machine, but that thundering anachronism has no place in the modern world. Its competitors are long dead.

The Thema, however, has plenty of competitors, most of them from Germany – M5s and the like. It's a bizarre cocktail, doubtless mixed by the corporate power brokers at Fiat. There have been Ferrari-engined Lancias before, most recently the Stratos, even Ferrari-engined Fiats – remember the beautiful, curvy Fiat Dino? – but never anything as seemingly crass as this. On the face of it, the 8.32 (eight cylinders, 32 valves) looks like a product of the British industry in its heyday, when all sorts of evil combinations of body and engine were concocted and foisted on innocent dealers.

The Thema deserves a sympathetic hearing, however. Lancia hasn't merely taken a Ferrari V8, cold, and shovelled it under the bonnet unaltered. Instead, the Maranello engineers have refined it to suit its new clothes, principally by tooling up a new crank, with throws of 90deg rather than 180, the better to suit its less frenetic character. And the rest of the car has been judiciously titivated. The ambience of the cabin has been altered as much as possible without incurring massive tooling bills. The suspension has been upgraded (and in this latest model, equipped with electronic damping), the brakes adapted for ABS, new Ferrari-style five-spoke wheels cast, and an outrageous fold-away spoiler built into the boot lid.

As a measure of its seriousness, Lancia assembles 8.32s away from the bustle of the main Thema production line, at the San Paolo plant where alleged craftsmen build the Lancia Ferrari at the rate of five to 10 a day. Why 'alleged'? Because there were not a few faults in the car we drove, most of which would be unacceptable in a £5000 Uno. This machine costs £37,500. Still, to keep the record straight, we'll add that the £48,102 Mondial hardly behaved impeccably either. But that comes later.

Aside from the fact that these cars are, in their way, four-seat Ferraris, and worthy of comparison for that reason alone, we had another question to settle. We all know that a well-driven Golf GTi can hold station with a Ferrari on today's coagulating roads. That's because GTis and their ilk are spectacularly quick and competent, but it also has a lot to do with size. Supercars (911s excepted) are just too big to funnel down B-roads and byways at speed – there isn't room enough left for oncoming cars. GTis, on the other hand, can usually squeeze by.

The Thema, however, because it's a bulky executive machine that cannot claim the size advantage of a Golf, must find other ways to outpoint the Mondial. So, the question is: Has a modern executive car got the dynamic nous to outgun a Ferrari?

We went to Exmoor to find out. Fast open roads, tight, tangled secondaries, and a fast motorway to start and finish with were enough to give us the answer.

Sitting in a supercar is not like sitting in ordinary machinery. And sitting in a Mondial is not quite like sitting in other supercars. From the driver's seat, the cabin seems curiously wide in relation to its length, and consequently spacious. Yet there isn't vast room for legs and feet, chiefly because the footwells are invaded by bulky wheel-arches. You also sit well forward – the B-post is some way aft of your head – and there's a lot of space between you and the ceiling, so that you look absurdly small, almost lost, inside the car. The reason is the Mondial's bizarre architecture, the result of efforts to turn the mid-engine two-seat 328 into a four-seater. More length clearly had to go into the equation,

Mondial/Thema battle fought on Exmoor. Ferrari much sharper handler, can be tempted into easily caught oversteer

'In the Mondi
you're aware of the
sound of reciprocatin
engine parts'

but height was needed, too, because the rear seats, being mounted over the fuel tanks, are set higher than those in front, enforcing the taller roofline. Given all these constraints, it's remarkable that Pininfarina made the Mondial look as good as it does. But we wonder whether its relatively poor secondhand value hasn't got a lot to do with the slightly ludicrous driving position – it can't be a good car for poseurs.

The tall roofline does make the car airy, though, a feeling reinforced by the void created by the broad centre console separating you and front-seat passenger. There won't be quite the same distance between you and anyone in the back – they'll be breathing down your neck, doubtless with increasing agitation, as they struggle to find comfortable stowage for knees and feet. Their quest will be in vain, though, because there simply ain't the space. This car will not seat four, even if there are four seats.

The Mondial driver doesn't sit conventionally, either. The pedals are set well over to the centre of the car – those damn' wheel-arches again – so that your legs stretch to the left. The steering column, conversely, points away to the right, leaving the wheel hanging at a slightly odd angle.

But, for all that, the car isn't uncomfortable, presumably because the seats are well-sculpted. Not that they look it. Lavishly upholstered in matt vellum leather, they nevertheless contrive to look mean because they are neither broad nor long. Nor do they offer much in the way of adjustment. If you're short, well, you'll just have to crane.

The Momo wheel can at least be adjusted, for rake and reach, and it's freed by an unusually lustrous chrome handle, one of only two plated parts in the cabin. The thin-stemmed gearlever is the only other chromed item – topped by a black spherical knob, it resembles a dinner gong stick.

If you turn on the Ferrari's ignition, but don't fire up, you'll hear the frantic whirr of a fuel pump priming for action. But ignition isn't that explosive – the V8 seems quite civilised. No loud-mouthed histrionics here.

You'll probably have some trouble engaging first before moving off. It's not that the dog-leg position is awkward – simply that, when the oil is cold, the lever is recalcitrant. Getting into second is difficult, too. At least the clutch is easy – light and well-damped.

Move away, and you're soon aware of a smooth stream of eager power, though you won't be awestruck by its intensity. The Mondial may be potent – there are 270 horses on call – but it isn't the lightest sports car around, and the poundage dulls the edge of eagerness. In any case, the engine's not warmed up yet, so we're limiting revs to 3500rpm. Even so, performance is brisk, not least because the ratios are stacked close, and the overall gearing is low – fifth is a puller, not a cruiser. That means there's entertainment off the motorway, but a fair bit of noise on it.

You hear different sounds in a Ferrari. The thrash of valves, belts and cams isn't far away, and you're more conscious of reciprocating parts than you are in other cars. It's something to be savoured for many miles, but not four-hour motorway journeys, and not every day.

Motorways are the Lancia's speciality. It'll devour them like a fat Italian eats spaghetti. Noise is not something the 8.32 makes much of, at least when cruising. But take it away from the nation's arteries, and it'll show you a dual personality, turning from tabby to tiger at the stretch of a toe.

In spite of its somewhat prosaic looks – the Thema was conceived as an elegant first-class carriage, not a sportsman's toy – it manages to convey its exotic pretentions as soon as you open the door. Your eyes and nose register one thing – leather, hectares of it. It swathes the seats, the dashboard, the door tops, even some of the ceiling. There's walnut, too, though it doesn't look quite as opulent, somehow, because it has a matt finish.

The steering wheel is swaddled in leather, too, and thoughtfully dimensioned for the eager driver. He'll enjoy the sight ahead of the wheel as well – a large speedo and tachometer, and six supplementary gauges whose messages range from oil temperature to whether the injection system is functioning properly. The markings are yellow on black.

The Thema driver will discover nothing strange in his seating position. Pedals, gearlever and wheel are all where they should be (except that they're on the wrong side, of course), and the seat itself is amply proportioned and well shaped. Adjustment is electric, and there's a heating element to take the chill off cold leather.

Turn the key, and it's not the dramatic buzz of fuel pumps that you hear, but the sound of servos adjusting the Saab heater (the controls have been lifted straight from the 9000) as it prepares to embalm you in a thermostatically controlled environment. Fire up, and once the idle has settled, you'll hardly notice the engine turning. Nor does it make much noise if you run it to only 3500rpm. Confine the engine to these speeds, and you'd think the Lancia Ferrari tame, a bit of a disappointment.

Westbound on the M4 come hints that there's more to it than that. When we slide

Thema is a cumbersome handler compared with pukka Ferrari: much more body roll. Understeer noticeable

'The engineers have refined Ferrari's V8 to suit its Thema clothes'

down slip roads, or accelerate in lower gears to clear dawdling traffic, the V8 lets rip a cultured growl. No doubt about it, we think. One of the greatest pleasures to be had from a Thema Ferrari is aural.

M4 and M5 done, we wind towards Porlock Weir, our overnight stop on the north Somerset coast. In the dying light of the sun we encounter a Golf giving chase to a Toyota. The pair are darting along the A358, a road that nuzzles the Quantocks as it heads north-east for the Bristol Channel. In the Mondial now, we naively think that when a long straight comes up, it'll simply be a matter of dropping to third, tickling the throttle and spearing by.

Only it never happens. The odd straight appears, but these two hatchbacks are well driven by a duo who seem to know the road and have doubtless been goaded on by the sight in their rear-view mirrors.

The Ferrari feels quick, but not quick enough to mount a successful charge. And the Golf wasn't even a GTi. We arrive at Porlock Weir sprouting doubts. Are today's wheel-clad appliances this close to a Ferrari? Shouldn't the Mondial's performance advantage feel absolutely decisive? Or has our driving turned tame?

'It's like a zoo up here,' exclaims snapper Wren as we erupt onto Exmoor at 5am the following day. Rabbits scuttle from the road, squirrels dart for the verge, horses and foals forage shoulder to shoulder with sheep, deer shy away nervously. This is where the animals take revenge on the motor car, and leave roadside presents for tyres and sills. There's a fair bit of low speed dodging and weaving – we don't fancy car washing at this hour, and we certainly don't want to cull the indigenous population.

In spite of this, it doesn't take long to discover that the Mondial has found the arena in which to put on an exhibition performance. And while it does it, we enjoy sensations quite alien to anyone used to ordinary transport. We're also presented with a challenge, because this isn't the easiest car to punt with speed and finesse. It demands a healthy degree of concentration.

If you can string five bends together, and go up and down the gearbox in a honeyed blur, and position the car correctly through a curve, and don't brake too abruptly and late, and perhaps tease the tail out a shade as you leave one bend behind before hurtling into the next, you'll feel good. If you get it wrong, well, you won't be punished. The Mondial is not a cantankerous car. But there's a huge difference between playing the piano with technical competence and playing it with bravura and emotion. Yet that is how the Mondial should be driven. You won't get the best from it if you don't, you may not even understand all the fuss about Ferrari.

Both cars use Ferrari's glorious V8. Thema (above right) has 215bhp, Mondial (right) gets 270. Both need revs

'The Thema was conceived as an elegant carriage, not a sporting toy

After the Mondial's virtuoso performance, the Thema, it seems, would be better left waiting in the wings, possibly for Godot. It was born as cosseting transport, and doesn't easily take to the role of delivering tactile excitement. The steering is light and seemingly reluctant to talk, the suspension soft and unrevealing, the car apparently too big and cumbersome to be a handy ally in a tight spot. You're physically further from the action, too, sitting several telling inches higher above the road than in the Ferrari.

One thing the Lancia does well, though, contrary to all predictions when it was launched, is put its power down. Those front wheels have 215bhp to introduce to the tarmac, and unless you're foolishly clumsy with the throttle, they'll carry out their task with some acumen. Torque steer isn't totally absent, but it certainly isn't a problem – there'll be no white knuckle moments if you accelerate venomously on crusty tarmac. Spinning front wheels aren't something you encounter, either, unless the roads are sodden, and in those conditions, any potent beast should be handled with circumspection. But when the roads are dry, you'll find good adhesion unless you go mad.

So, you can move with alacrity in the Thema. And you damn' well should be able to. What you can't do is hurl the car through the tight, rubber-crushing bends that make up most of Exmoor's roads. Enter one of these too fast and the unequivocal message is that you're barely in control of a large mass that does not wish to change direction at the pace required, thank you. Of course, it's no problem to get it around the turn. Off with the throttle (at least there are no tail slide dramas, unless the speeds are crazy), on with more lock, and here's hoping you aren't scraping too much expensive rubber off those 205/55 Goodyear Eagles. This isn't a very satisfying way to get about. You're seduced by a wonderful engine, then made to take a metaphorical cold shower by a chassis that doesn't want to play.

Happily the Thema is better on faster, less sinuous roads. The understeer diminishes, the feeling that you're swinging a pendulous mass from corner to corner evaporates, and you can actually power through turns, and with some gusto.

That's when you really enjoy the engine, which has to be one of the smoothest, sweetest-sounding motors you can buy today. It sounds so good, you think it must run in molasses. It's not until the yellow tachometer needle has swung past 4000rpm that you're convinced of the engine's potency. The boast is that 80percent of the available 210lb ft of torque is on hand from as little as 2500rpm (the rest of it is released at 4500rpm) but it takes a decisive prod at the accelerator to stir the V8 at low to middling revs. Which is one reason why torque steer and wheelspin aren't the problem that they might be. Beyond 4000rpm, all the way to the 7000rpm red-line, the Thema takes on a new, aggressive personality and charges for freedom, engine wailing. When that V8's revving hard, you start to think that if there were money to burn, you'd buy a Thema Ferrari just for the noise it makes. You won't

Ferrari a little noisy on long trips. Thema more the executive express. Fwd Thema (below) has very little torque steer

'One thing the Lancia does well, contrary to predictions, is put power down'

hear the same symphony in the Mondial, because that has a different crank, whose 180deg throws make the engine sound *corsa*.

The Mondial is no less exhilarating for all that. You can pick out more individual sounds, and, in time, probably identify them. A riding mechanic from days of yore would doubtless be able to inform you of the

engine's condition from the passenger seat: 'I think she's running a bit lean on the second cylinder bank, m'lud.' You can hear the gearbox, too, its high-pitched whine rising and falling in concert with your right foot. It may get wearing on motorways, but on winding roads it's wonderful, part of the Ferrari dream.

And it gets you in the mood. When you and the car are in full cry you could be in the Mille Miglia. Or the Targa Florio. Or on the Futa pass. More likely you'll be on the A38, but the excitement will hardly diminish.

Much of it stems from the acceleration of the car. Oddly enough, the Mondial and Thema are pretty evenly matched when it comes to performance, in spite of big differences in output and gearing. The 270bhp Mondial will sprint to 60mph in 6.4sec and on to 150mph, yet the Thema, whose engine musters 55bhp less at 215bhp, reaches 60mph in 7.1sec and tops 149mph. All slightly baffling.

There's not much in it on the road. If the Ferrari gets ahead, it will be because the roads twist violently, and the Thema's chassis can't cope. Like the Thema, the Mondial suffers from a paucity of low-down torque. Of course, we're talking in relative terms here. The effect is heightened only by the ferocious acceleration that sets in once 3800rpm has disappeared, and power emerges in torrents. But the real flood is reserved for those who venture beyond 6000rpm and climb to the 7000rpm peak. If you're to travel blindingly fast, the tacho needle must hover here. Which is why we weren't confident of whipping past that Golf – we hadn't yet fully grasped how hard the engine had to be kept spinning to extract full commitment from it.

Once this is learned, you're beginning to unravel the secrets of Ferrari driving. Keep the engine close to its wailing crescendo, enjoy the clock-clack as the gearlever thrusts from gate to gate (it sounds like you're cocking of a rifle at every shift), the siren whine of gears meshing against a base line of rumbling tyres. By the time you've struck the red-line a few times your blood will be up, and you'll be exploring the chassis.

It seems a pretty tame ally at first. The steering may be beautifully measured in its response, but it sure isn't quick, and nor is the turn-in. An Esprit feels more agile. There's even a trace of understeer. You can expunge that, though, with an enthusiastic stab at the throttle. The result will be deliciously neutral handling, or if you've been over-zealous, a dose of easily quelled oversteer. Once you've suppressed the thought that you're dicing with £48,000, it becomes intoxicating, and the speeds build. And as you travel faster, you're grateful that the steering isn't over-eager – progress would turn nervous otherwise.

Not that there's cause to worry about the Ferrari's roadholding. Grip from the generous tyres is prodigious, and the Mondial doesn't suffer the sudden heart-stopping breakaway that sours some mid-engined machines. Instead, the rear tyres slide gently and tidily, so that you've plenty of time to rein them in. If they didn't, you'd wish the steering were quicker. The steering itself is equally well-mannered. There's some kickback, but only enough to let you know that you're on a road that may soon need work. But it's as well to keep a firm grip on rim, because it can writhe noticeably under braking. But then the Mondial, like any good Ferrari, is a car that needs commitment. Provide that, and you're guaranteed a good time. Its brakes match that disposition. You need to work, to push them hard. But the rewards are precision and ample stopping power. The Lancia's, also ABS-equipped, are softer and mushier, but still effective. Like the Mondial's brakes, they are in character.

The Thema doesn't demand much commitment at all, if you don't feel in the mood. You can waft along, thoughts miles away, as you can in any executive car. But in the right circumstances, the 8.32 can be an invigorating partner. Long sweeping bends are its forte, bends where understeer won't kill progress. The trick is to throw the car in, and fight through the mushy responses that Fiat's new electronic suspension seems to send. Practise the technique with enough zeal, and you'll be surprised to discover that the Thema actually turns neutral, plunging through turns in a pretty satisfying manner.

But that new suspension has done some damage to the 8.32. It still uses MacPherson struts all round (there's none of the Mondial's

Mondial has strange driving position, a legacy of trying to cram in a couple of rear chairs. Seats comfortable. Rear cramped

'The Ferrari should be driven with bravura and with emotion'

ourity here; that car has wishbones at each corner), but electronic dampers have been added to the equation. They appear sophisticated in theory – there are accelerometers and a variety of sensors, sending messages to the computer that disciplines the dampers, but its orders seem confused. One minute the 8.32 travels with a suppleness approaching a Citroen's (though it has to be said that you'll make the analogy rarely), the next it goes to pieces, displaying all the damping control of a Datsun Stanza.

That the driver has two settings to play with – sport and auto – doesn't help matters much. Even in auto, the ride is never particularly good, often jigging and crashing. In sport, the symptoms simply worsen, without any apparent improvement in roll control or poise. This is a system that needs a lot more work.

The Thema's case isn't helped by the quality of its body, which like other Type Four cars (Alfa 164 excepted) doesn't seem stiff enough. Especially with a bigger

engine and a long list of extra equipment dragging it down. Potholes have the body shaking and jarring, the facia and fittings singing a twittering accompaniment, to the extent that you find yourself steering around ruts. Ride is easily the 8.32's most disappointing feature. A Jaguar is light years ahead. The sad thing is that if Fiat hadn't meddled with the original passive suspension, the car would be a whole lot better than it is now; more the car we've admired in the past.

Most amazing, the Ferrari completely shades the Lancia for ride. It not only has far better damping control, though that's a hollow victory against the 8.32, but also delivers a more comfortable ride, even at crawling speeds. And all this is achieved without muting the car's messages to its driver. The Mondial has another advantage, too – it's body is tremendously stiff. The car feels as one even on terrible roads – there's not a creak or groan to be heard. You soon feel so confident in its build that you believe it could be dropped from 20feet and survive.

Only our car didn't. A mere 320miles after we had collected it, the engine sprang an oil leak which gradually coated one half of the engine bay, then worked its way onto the rear screen. When we stopped, wisps of smoke drifted from the engine bay. Imagining the most expensive conflagration Exmoor has seen since World War Two, we decided to abandon the Ferrari at the Anchor and Ship Hotel in Porlock, our friendly refuge the previous night, for Maranello Concessionaires to collect.

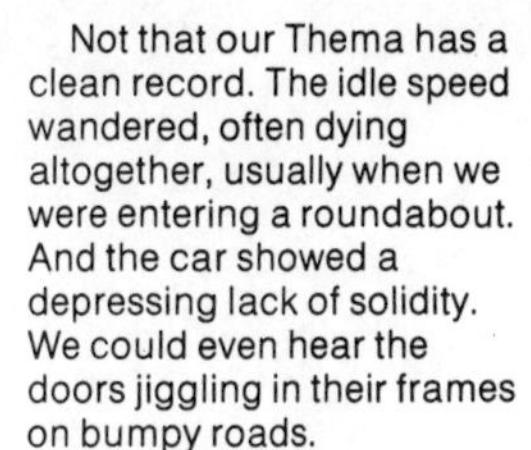

Not that our Thema has a clean record. The idle speed wandered, often dying altogether, usually when we were entering a roundabout. And the car showed a depressing lack of solidity. We could even hear the doors jiggling in their frames on bumpy roads.

We left the Thema unconvinced that it will be reliable, but of the Ferrari, apparently so beautifully built, we are prepared to put the leak down to bad luck. It shouldn't happen, mind, in cars of this price.

As to which is the more satisfying, there can be no

doubt. The Ferrari is infinitely more entertaining than the Lancia, scoring decisive advantages in every dynamic department other than performance. And we know who was responsible for that. Ironically, its biggest failing is that it cannot really carry four people, the very thing Ferrari set out to achieve. Even children would complain at incarceration in the rear.

The Thema will carry four, even five with ease, but what it cannot do is double as Ferrari and family saloon. It has the engine, but not the chassis. A stiffer shell, four-wheel drive and conventional steel suspension might give the Mondial a stiffer challenge, but such things we're unlikely to see. The Ferrari remains unchallenged, by the executive class at least. And, magnificent though the Thema is, that has to be good news for the supercar.

Thema awash with cow hide. Comfort hurt by poor ride. Left-hand drive only. Bad oil leak brought Mondial to a halt (above)

'The Thema's chassis just can't cope, when the road twists violently'

MONDIALt

Little t, more cc

BY PAUL FRERE

PHOTOS BY GUY MANGIAMELE

THE FERRARI MONDIAL t is more than a new model. It is the spearhead of a new generation of Ferrari V-8 models, not only more powerful, but also more technically advanced. There is even better handling and more comfort than ever before. The new Mondial t has its engine mounted longitudinally and is hardly distinguishable externally from its predecessor which had a transverse engine. The clever arrangement of the transmission has even made it possible to retain the same 104.3-in. wheelbase as before, though this is pretty long for a so-called 4-seat car with rear seats strictly for dogs.

To save space, the longitudinally mounted V-8 is mated to a completely new transverse gearbox, an arrangement previously used in Ferrari's Formula 1 cars until the team switched from normally aspirated 12-cylinder engines to shorter turbocharged V-6s. A twin-plate clutch is mounted overhanging the rear axle and is driven through a long shaft, which locates the final drive as close as possible to the rear of the engine. From the clutch, a pair of gears sends the drive forward to a pair of bevel gears, driving a transverse-mounted gearbox containing the 40-percent limited-slip differential of the final drive, only a few inches to the rear of the cylinder block. This, together with a switch to dry-sump lubrication, lowers the engine about 5 in., further lowering the car's center of gravity by 1.4 in.

The 3.4-liter 32-valve V-8 is a development of the well-known 3.2-liter and retains the 180-degree ("flat") crankshaft characteristic of V-8 racing engines. This configuration gives pride of place to an efficient exhaust system at the expense of perfect balance, because engine forces of the second order remain unbalanced. The engine is well insulated from the body, and vibration-free, but a high-pitched scream is emitted as the superbly revving engine approaches its 7500-rpm redline.

The new engine develops 296 bhp DIN at 7200 rpm and a maximum torque of 238 lb-ft at 4200 rpm benefits from the latest Bosch Motronic 2.5 fuel-injection system with a hot-wire air flow meter. Thus equipped, it runs a 10.4:1 compression ratio, even though it is tuned to use unleaded fuel of only 95 octane rating.

But the more powerful engine and the completely new transmission are not the only novel features—as far as Ferrari is concerned—in the externally almost unchanged car. The Mondial t is the first Ferrari featuring power-assisted rack-and-pinion steering. Also a Ferrari first are the Mondial t's electronically controlled, variable stiffness shock absorbers with three driver-selectable ranges.

Helped by the long wheelbase, the soft setting makes the car quite comfortable on rough roads and at leisurely speeds, while the medium setting provides a good compromise between comfort and the tautness required for fast driving on most types of roads. It is the setting I would choose most of the time for my many long drives across Europe, while the hard mode is just what you want for the occasional excursion on the race track so many Ferrari owners enjoy. It was just what was required for some quick laps of Ferrari's own Fiorano circuit.

The new power-assisted steering makes an enormous difference to the driveability of the car. I am not a fanatic of power-assisted steering in sports

cars, but neither have I liked most of Ferrari's manual steering systems. Even rack-and-pinion steering from Ferrari has always had a somewhat dead feel and some friction. The Mondial's power-assisted steering does away with all this. It is not very sophisticated, but it is smooth, accurate, not too light at speed and provides sufficient feedback.

Its gearing is also quicker than the old Mondial's, enabling the driver to perform quick corrections if necessary. When exploring the limits, the necessity arises more often than perhaps it should, because of the Mondial t's propensity for final oversteer.

The way the front end is glued to the road and its ability to turn in at the slightest movement of the steering wheel is wonderful. Never is any understeer felt and there is never any fight to hit a bend's apex. Nevertheless, the development engineers have slightly overshot their target. While the front end remains glued to the asphalt, the ultimate cornering speed, though still high, is limited by the early breakaway of the rear end, even with moderate power applied. This should not be confused with power oversteer which, with 300 bhp on tap, is easy to obtain in the appropriate gear.

And thanks to the limited-slip differential that prevents the inside wheel from spinning wildly while the outside one stubbornly continues to grip, power oversteer is easy to control. If I were responsible for handling at Ferrari, I would specify a slightly stronger front anti-roll bar, maybe in conjunction with a lighter rear one to achieve more neutral handling in the t.

THE BRAKES stood up to the Fiorano test with fine endurance. True, I made a point of braking rather early and smoothly, but Fiorano is an extremely demanding course on the brakes and taking this into consideration, their performance was very creditable, with no sign of fade. The ATE anti-lock system is now standard equipment and proved very effective in obtaining a 1g average deceleration from 60 mph without any deviation from the set course.

With its new 3.4-liter engine, the new Mondial t is now a full match for the Ferrari 328, but its engine is not only a bag of power. It is also utterly flexible. Not only does it idle like clockwork, but it

will also pull away from 1000 rpm in 4th or even 5th gear without hesitation and starts pulling quite strongly from around 3000 rpm on. The gear ratios themselves are quite well chosen, with a very vigorous 3rd reaching up to just around 100 mph. Though the gearbox is new, the cable-operated gearshift mechanism remains typically Ferrari: stiff with a lot of perceptible friction and slow when shifting dogleg-wise from 1st to 2nd. The limited-slip differential is very effective when accelerating out of a slow bend in a low gear, but can be felt creating some roughness when rounding a tight bend under moderate acceleration.

Air conditioning with automatic temperature control is standard, and air output is individually and separately adjustable for the driver's and passenger's sides. While the Mondial t test car was a not-too-well kept demonstrator, the new cars I examined on the assembly lines were beautifully finished and suggested excellent workmanship.

It may not provide the fun and exhilaration of an F40, but the Mondial t is a jolly fine, very fast car which inevitably makes one think of other things to come. I just cannot wait for the 348 and I am bound to dream, one of these days, of power-assisted steering in a Testarossa . . .

ROAD TEST DATA

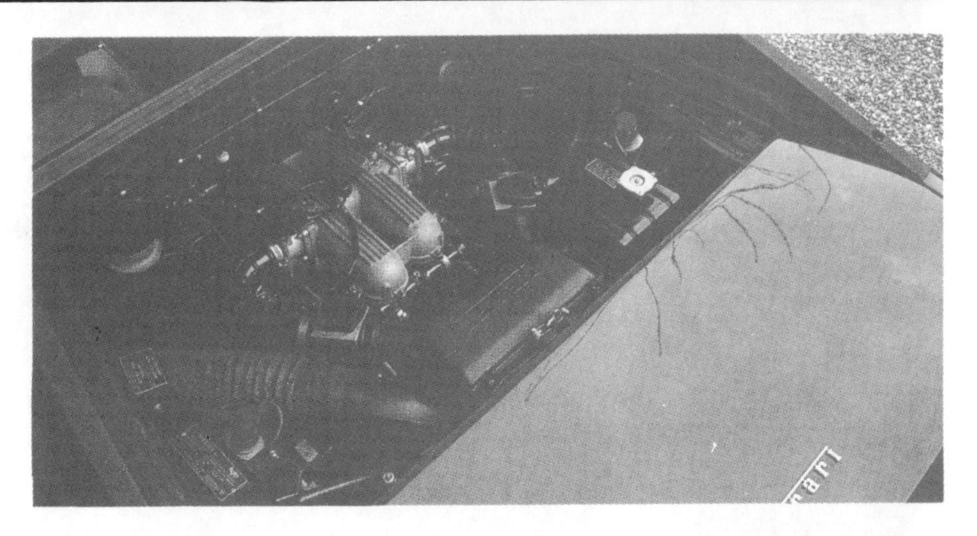

GENERAL

Wheelbase, in.	**104.3**
Track, f/r	**59.9/61.4**
Length	**178.5**
Width	**71.2**
Height	**48.6**
Trunk space, cu ft	**5.0**
Fuel capacity, U.S. gal.	**22.7**

CHASSIS & BODY

Layout	mid engine/rear drive
Body/frame	steel/tubular steel
Brake system, f/r	11.4-in. vented discs front and rear, ABS, vacuum assist
Wheels	cast alloy, 16 x 7 front, 16 x 8 rear
Tires	Goodyear Eagle VR; 205/55VR-16 front, 225/55 VR-16 rear
Steering type	rack & pinion, power assist
Turns, lock to lock	2.9
Turning circle, ft	39.0

Suspension, f/r: unequal-length A-arms, coil springs, tube shocks, anti-roll bar/unequal-length A-arms, coil springs, tube shocks, anti-roll bar

ENGINE

Type	**dohc 4-valve V-8**
Bore x stroke, mm	**85.0 x 75.0**
Displacement, cc	**3405**
Compression ratio	**10.4:1**
Maximum engine rpm	**7400**
Bhp @ rpm, DIN	**296 @ 7200**
Torque @ rpm, lb-ft	**238 @ 4200**
Fuel injection	elect. port
Fuel requirement	95 pump oct

DRIVETRAIN

Transmission	**5-sp manual**
Gear ratios: 5th	**0.86:1**
4th	**1.09:1**
3rd	**1.15:1**
2nd	**2.11:1**
1st	**3.21:1**

ACCELERATION

Time to distance, sec:	
0–1320 ft (¼ mi)	**14.2**
Time to speed, sec:	
0–40 mph	**3.3**
0–50 mph	**4.6**
0–60 mph	**5.9**
0–70 mph	**7.7**
0–80 mph	**9.6**
0–90 mph	**11.8**
0–100 mph	**14.9**
0–110 mph	**18.7**

SASKIA VAN STEDEREN

Done to a t

The new Ferrari Mondial t Cabriolet is set to capture the public imagination in a way its predecessor never could. With a bigger, longitudinally-mounted V8, Ferrari's only true convertible is now the dream car it always promised to be. Peter Robinson reports

Discreet changes to nose and tail and smaller side intakes distinguish new Mondial. Seats are far more supportive but chrome gear lever takes some getting used to. Adjustable Momo steering wheel is well placed. Engine gives smooth responsive flow of power through incredibly wide rev range

Of all current Ferraris, the Mondial is the least emotive. Deprived of the flawless beauty of the 328, the stirring presence of the Testarossa, the shattering potential of the F40, even the gracious manners of the enduring 412, the Mondial has always been something of an uncomfortable compromise. Understated, the styling lacks the strong personality expected of a Ferrari — a reaction that driving only served to verify.

Creating an attractive body around mid-engined packaging that insists upon two-plus-two seating has never been easy. Bertone tried with the 308 GT4 in 1973, only to see his efforts completely overshadowed by Pininfarina's two-seater 308 GTB two years later. Yet Pininfarina's attempt at a successor to the GT4 resulted in the rather sombre Mondial. And for all its greater length and surprising height, the Mondial was no more than a mediocre two-plus-two. Moreover, with its bulk came weight so the performance was ordinary, at least by the standards demanded of a Ferrari.

I remember writing, in 1982, "downright slow . . . intolerably uncomfortable . . . a model that is not destined to become an instant Ferrari classic," and getting into dreadful trouble with the men from Maranello who told me it was the worst road test ever penned of a Ferrari. Sales figures seemed to bear out my judgement.

That was before the Quattrovalvole heads and the 3.2-litre V8. But although the increased power restored the performance to an acceptable degree, it did little to lift sales. It was left to the release of a convertible version — Ferrari's only truly open car — in 1983, and initially for the US market only, to bring some respectability to the Mondial's sales figures.

In 1988 Ferrari delivered 511 Mondials, 219 of them Cabriolets. On the other hand, 2242 GTB/GTS models were sold. In the UK, Cabriolet sales of 17 were just one shy of matching the coupe, while in the warmer climes of the US Cabriolet sales were almost five times those of the coupe. Practical, four-seater Ferraris evidently lack the charisma of the two-seat models, a fact of life the beautiful new 348 is certainly not going to alter.

That's a shame, for the Mondial has been developed painstakingly into a fine car, the latest 't' modifications bringing improved performance, handling and comfort levels and even subtle changes to benefit its appearance. In convertible form, especially in convertible form when spring flourishes sunny and bright and an open Ferrari beckons, the Mondial t merits close scrutiny.

We arrived in Maranello expecting to drive the new coupe but, not for the first time, were told there had been a change of plan. Instead of the coupe we were handed the keys to a Prova Cabriolet and left in the care of one of the legendary Ferrari test drivers. Claudio Ori, he of the abundant moustache, shiny bald head and the soft suede shoes that are the trade-mark of Italians who like to drive fast cars fast, would drive the car outside the plant for photography. Our driving was to be limited to the tight Pista Fiorano circuit, the insurance problems of an outsider driving on Prova plates being simplified on Ferrari property.

Ori's skill behind the wheel isn't matched by a talent for lowering the soft top of the Cabriolet. If there is a quick and simple method we were unable to find it, so there are no photographs here of the Mondial with the roof in place. Once down, we weren't about to raise it again. Take it as read that the car looks far more alluring with the roof tucked away under the still-exposed tonneau cover, despite the large black cover being all too conspicuous against the silver body.

The new model doesn't look significantly different, but under that lightly modified body — the most obvious change being smaller side intakes, though there are also discreet alterations to the nose and tail that reduce the drag co-efficient to a still poor 0.40 — the larger capacity V8 engine now sits longitudinally, running at 90deg to the crankshaft.

If there is a finer automotive engine in production in the world than the 3.4-litre Ferrari V8, I know not from where it comes

This is fundamentally the same powertrain unit that will appear later this year in the 348. Ferrari says the change is to promote serviceability, but primarily to allow the engine to be lowered in order to drop the centre of gravity and therefore further improve handling. On the old Mondial (and 328) the gearbox lay beneath the engine; fitting the gearbox at the end of the engine would have created too long a drivetrain. The solution chosen by Ferrari's engineers was adopted from the 312T Formula 1 cars of the late '70s. The result is that the engine is now five inches closer to the blacktop.

To enhance performance, the capacity of the double overhead cam V8 has been taken out to 3405cc by increasing both the bore and stroke, up from 83mm x 73.6mm to 85mm x 75mm. Maximum power has climbed to 300bhp at 7200rpm and torque of 238lb ft is developed at 4200rpm, compared with the 3.2-litre engine's 270bhp at 7000rpm and 224lb ft at 5500rpm. The gearbox is, of course, all-new with the drive coming from the engine via a 90deg bevel gear set to the hydraulically-operated clutch. This is mounted with the flywheel in an external housing at the end of the drivetrain, rather than being on the end of the crankshaft. A 40 per cent limited slip differential is coupled to the gearbox by cylindrical transfer gears. There's even a choice of final drive ratio, the standard ratio being 3.823 and giving 20.2mph per 1000rpm in fifth gear and a slightly higher 3.706 giving 20.8mph. Official Ferrari figures suggest a top speed of 158mph which represents 7800rpm — 300rpm into the red with the standard ratio — or 7600rpm with the optional ratio.

Watching Ori thread the Mondial through the traffic of Maranello, as we headed up into the foothills of the Apennine mountains to the west of the town, taught me much about the car. He drifted along, taking advantage of the engine's wonderful torque and rarely exceeding 3500rpm, slipping the delicate chrome gear lever in apparently measured changes from one ratio to another, pausing momentarily in neutral before pushing firmly into the next gear. Deftly, with the considerable skill that comes of daily exposure, Ori knows only the smoothest line through the corners, never appearing quick or even to be concentrating. Nonetheless we proceeded so swiftly as to leave any other traffic in futile pursuit.

It is a big car, the Mondial, long in wheelbase — at 104.3ins it's 11.8ins longer than the 328 — to make room for two small rear bucket seats. The cabin is well forward, with a short bonnet and steeply raked windscreen, and there are large fixed quarter windows, where the coupe has none, to provide additional rigidity. Restricted by the position of the engine, this is most definitely two-plus-two accommodation although, if the front seat occupants are prepared to compromise a little, it is certainly not uncomfortable in either compartment, though those behind will find wind turbulence excessive. Headroom, however, is not a problem.

Elastic and surprisingly comfortable, the ride quality is only spoiled over the most severe of bumps which induce some scuttle shake. ▶

New Mondial feels more agile and responsive, an indulgent car to drive with impeccable manners. Power steering is major improvement. Mondial looks more alluring with roof tucked away but lowering soft top proved a chore

◀ Body control is excellent, a three-position damper adjustment allowing the driver to choose between soft, intermediate and hard for the now Bilstein, rather than Koni, dampers. Ori selects soft for the transport stages, switching to hard only when really rapid driving is required for the camera. Knowing the dampers firm up automatically when vehicle speed or lateral forces build up, he prefers to leave the system on soft to enjoy the real benefits of the extra low-speed comfort.

Sitting low in the cabin on thinner yet more supportive front buckets, the rush of wind passing over our heads if not those behind, we whistle up the ridge top and zig-zag roads that have become familiar to two generations of journalists and a playground for numerous Ferrari test drivers to show off in the latest model. Here, where it has been tested and tested again, the Mondial's chassis balance and manoeuvrability are plain to see.

Ori's control is masterly and he slides the Cabriolet through the bumpy corners in long power drifts for the camera, spinning the car through 180deg from a standing start in less than its length to avoid having to drive down the road to find a turning space. You know he has done it all before and will do it again soon.

Suddenly it is lunch time and our photographer has finished for the morning. On the way back into Maranello, the sun glints off the Mondial's bonnet and heat is held in a

permanent shimmer above the vents to the engine. In an open car, you remember again, there's no necessity to travel quickly. There is so much pleasure in motoring gently, the senses absorbing the joy that comes from total involvement in the countryside. You smell the change of season, taste the perfect visibility, catch the wind in the Cyprus trees over the constant whirr from the Ferrari's engine.

Back at the legendary figure of eight Ferrari test track, an F40 circulates rapidly, thrilling the American visitors who've left their 30-year-old Mille Miglia chargers at the factory to ride beside another fearless Ferrari test driver. The F40 streams past the timing shed, reaching 150mph down the short straight. Claudio parks the Mondial beside an old farm building, now converted to a garage. It contains, along with a Fiat fire truck, a couple of pushbikes, a Fiat tractor and a pile of old racing tyres, five Ferrari F1 cars of recent vintages. Three carry the name Alboreto, one Berger and the fifth and oldest, Villeneuve. The F40 disappears to be replaced by a Testarossa with Leitz Correvit performance testing equipment.

Then it is our turn. Two hours alone on Fiorano, to play at being Nigel Mansell in a Mondial convertible. There might be better ways of spending an afternoon in Italy, though for the moment I can't think of any.

You drop down into the Mondial feet first, using the windscreen pillar for support, because the cushion is well forward and the doors don't open very wide. The pedals are still offset, due to the intrusive wheel arches, but you're only aware of this for a brief moment. The Momo wheel is adjustable for height and reach and is now straight ahead and very comfortably positioned at a more vertical angle than on previous Ferraris. The excellent driving position is enhanced by properly comfortable bucket seats that have clearly been completely redesigned and provide good lateral and thigh support. The controls are conveniently laid out, mostly on the central console, while the instruments — the tacho is graduated to an impressive 10,000rpm and the speedometer to 280km/h (174mph) — are easy to read under their heavily shrouded cover. The interior is surprisingly conventional and spacious, and beautifully finished in soft leather. Only the traditionally delicate-in-appearance-but-not-in-action chrome Ferrari gear lever, feeding into the exposed gate, requires some intimacy before shifts become second nature.

Ori had been having trouble starting the Mondial all morning, the engine demanding frequent cranking, a problem he explained isn't normal on this early prototype. Fortunately it fires immediately, the V8 idling quickly even when warm. The clutch is heavy yet flowing in its movements, the gearchange less so; it needs to be teased out of one ratio and forced into another before you understand that a firm hand is required. And, as always with a Ferrari, there is a tendency to look down at the gate when changing gears to ensure you have selected the right ratio. The indecision soon fades. First is down to the left and reverse opposite, the change being heavily spring-loaded to the second-third plane.

If Ferrari allowed its low-emission engines from the early '80s to fall behind established performance standards, the latest engine, tuned to run on unleaded fuel, restores the status quo. If there is a finer automotive engine in production in the world I know not from where it comes. There's torque enough at 1000rpm to pull fifth gear and yet it will run effortlessly to the red line of 7500rpm, the deep gutsy ring of the V8 building from 2000rpm with a wonderful visceral induction roar that expresses the engine's power better than any performance figures.

This engine delivers a smooth and responsive flow of power right through its incredibly wide rev range. Only over the last 500rpm to the red line does it begin to display any sign of stress and that is more a shrill gear whine than inherent harshness. So progressive and expansive is the spread of power that the Mondial is a docile car to drive, tolerant of infrequent gear changes, demanding little of the driver once the gear shift has been perfected.

Two hours on Fiorano to play at being Nigel Mansell in a Mondial convertible. What better way to spend an afternoon in Italy?

So tractable is the engine that first can be engaged with 1500rpm on the tacho, the car then moving forward on idle. No, Ferraris aren't intended to be driven in such a way. Push down firmly and the car sprints forward, the engine shattering the quiet. Ferrari claims a 0-62mph time of 6.3secs for the Mondial — 1.1secs faster than the old 3.2-litre version — and that seems about right, though it happens remarkably effortlessly.

Two things strike you immediately when driving the Mondial. The first is that the power steering, fitted for the first time on a Mondial, is so light at low speeds as to make you doubt its competence. But turn into the first corner and you begin to appreciate its accuracy and sensitivity. Beautifully weighted, with 2.9 turns lock to lock compared to the previous model's 3.4, the power steering is one of the more obvious reasons why this Mondial is so superior to the old. The old rack and pinion kick-back that afflicted former models has disappeared, and the steering is so delicate that the car can be driven very quickly from the finger tips. No longer does understeer intrude on turning into a corner to take the edge off the old car's excellent handling. The new Mondial feels both more agile and more responsive.

There *is* gentle understeer, but even that is too strong a term for the way this Ferrari drives around corners. Yes, it is possible to kick the tail out on the tight Fiorano corners and hold the car in a lovely, balanced oversteer slide, if that's the way you want to drive. The grip provided by the Goodyear Eagle ZR55 tyres — 205/16s up front and 225/16s at the rear — is convincing, yet they are also progressive when they do lose adhesion. The new Mondial is an indulgent car to drive quickly, with impeccable manners.

The second aspect is the car's low gearing. Disguised by the engine's incredible flexibility and the breadth of its rev range, the low gearing means short ratios. First runs to just 40mph at the 7500rpm redline, second 62, third 89, fourth just 119 and fifth 151.5, although the engine will spin even faster for short bursts. Does the Mondial need a six-speed gearbox? It's a solution Ferrari is exploring.

Of course the immense torque of the engine means you don't have to be constantly changing gears so it is almost an academic point, exposed only when accelerating through the gears for the first time and before coming to terms with the span of the engine's range. Five minutes into the drive and any thoughts of low gearing are forgotten.

Today's Mondial has anti-lock brakes, the pedal communicating some sponginess that is an instant give-away, though there is little pulsating at the pedal when the anti-skid device comes into operation. Vented at each wheel and with floating calipers, the brakes are formidable.

We left the Mondial at Fiorano, wanting to prolong the drive and knowing that the lower-case t after its name has furnished the latest Ferrari with respectability at last. Practical, certainly, but also appropriately fast and with fine handling, the Mondial deserves serious consideration. And there's still the glorious attraction of the convertible body — a Ferrari convertible after all. ■

Maranello—

Ferrari prices are going through the roof. Testarossas fetch premiums for early delivery, F40s change hands for about $800,000, the last 412i's command ridiculous sums, and even the humble 328GTBs and GTSs are worth a small fortune now that their production is about to end. The only model that has remained virtually unaffected by the prancing-horse craze is the Mondial. Despite its virtues—good space utilization, elegant design, and well-balanced handling, to name only a few—*Ferraristi* have never been too fond of the practical two-plus-two. It is still too early to tell whether the most recent changes made to the body, the cabin, and the driveline will turn the wallflower into a commercial success. But our first test drive made it clear that the notoriously underrated Mondial is now the best-value-for-money Ferrari model—as well as a wonderful driving machine.

The "t" in the new model designation stands for transverse, despite the fact that the engine is now mounted north to south instead of east to west. We asked senior project engineer Franco Cimatti for an explanation. "You are right—the V-8 is installed longitudinally," says Cimatti, who acquired his neat West Coast accent during a seven-year stay in the United States. "But the placement of the transmission has changed from in-line to transverse. This configuration offers a variety of advantages. Among them are a lower center of gravity for improved handling, even exhaust temperatures on both cylinder banks for superior performance, cleaner emissions, and better reliability, as well as easier access for servicing."

Although there are still some on-paper similarities between the first-series Mondial and the new t model, the engine, the transmission, and the rear suspension have in fact been largely redesigned. The V-8 has been increased in capacity from 3.2 to 3.4 liters, with the maximum power output of the catalyst version boosted from 270 bhp at 7000 rpm to 300 bhp at 7200 rpm. Maximum torque is up from 224 pounds-feet at 5500 rpm to 237 pounds-feet at a more civilized 4200 rpm. The detail improvements made to the engine include a new Bosch Motronic management system, a more precise hot-wire airflow meter, an electronic ignition, narrower valve angles for improved thermodynamic efficiency, and a revised intake tract with separate plenums and a connecting valve that opens at full throttle. "Even the block is effectively brand-new," says Franco Cimatti. "It has been changed geometrically, and it features longer cylinder liners as well as a single timing belt, a more efficient water pump, dry-sump lubrication, a rerouted free-flow exhaust, and a lightweight alu-

BY GEORG KACHER

PHOTOGRAPHY BY PAUL DEBOIS

The capacity of the 32-valve V-8 has been increased to 3.4 liters. It produces 300 bhp at 7200 rpm.

The drivetrain is slimmer and approximately 3.9 inches shorter than the previous arrangement.

minum radiator. The power steering pump and the air conditioning unit are new, too."

The gearbox runs at a 90-degree angle to the crankshaft. This compact layout enabled the engineers to lower the driveline by about 5.1 inches, but at the same time, it required an unusually complex array of cog works. Torque is passed to the gearbox via a coaxial input shaft that drives first a helical gear and then a bevel gear before passing the drive on to the primary and secondary shafts of the transmission. This sounds complicated, but it works well. The two-mass hydraulically damped flywheel eliminates any undue vibrations, and the new cable-operated shifter prevents the gear lever from moving with the torque reversals. Since the clutch unit now sits at the end of the driveline, right in front of the rear bumper, it can be exchanged in about twenty minutes—a vast improvement over the old car, in which the same job took more than six hours.

The drivetrain of the Mondial t is considerably slimmer and approximately 3.9 inches shorter than the previous arrangement. The engine and the transmission are cradled in a new subframe, which entails repositioned mounting points, a slightly wider rear track, and mildly modified suspension kinematics. This installation is virtually identical to that of the upcoming Ferrari 348, to be launched at the Frankfurt Auto Show in September. The chassis was taken over unchanged from the original Mondial, but instead of open-profile C-section suspension arms, the 1989½ model uses lighter and more rigid box-section elements. To accommodate the standard anti-lock brakes, the scrub radius of the front suspension was reduced to almost zero. The standard tire size is 205/55ZR-16 in the front and 225/55ZR-16 in the back, but extra money will buy even fatter 220- and 240-section gumballs.

The cosmetic revisions are inconspicuous and subtle. The Mondial t has mildly flared instead of lipped wheel arches as well as flush-fitting body-color door handles and restyled side air intakes. Although these modifications do improve the car's aerodynamics, the drag figure is still a rather unexciting 0.40, down from 0.42. Progress is more obvious inside the t model, where we note wider footwells, better-positioned pedals, a smaller-diameter Momo steering wheel, more comfortable seats, higher quality switchgear, an uprated center console, and a brand-new instrument binnacle starring six round orange-on-black dials. Unfortunately, the air vents are still too small, the push buttons and the heat and ventilation controls are still an ergonomic mess, and the quality of the leatherwork is poor for a car that costs $82,000.

Before we hit the road, *ingegnere* Cimatti points out some of the less obvious alterations, like the new wiper blades with integrated washers, the power-assisted steering, the long-range homofocal headlamps, and a small switch next to the ungainly digital clock. "This rocker switch operates the adjustable shock absorbers," explains *signor* Cimatti. "The system, which is governed by vehicle speed and g-force, was developed together with Bilstein. It provides you with three different settings—soft, medium, and hard. The faster you go, the tauter the damping. Above 100 mph, the computer will automatically switch to the stiffest setting. At lower speeds, any brisk accelerating, braking, or cornering maneuver will alter the damper rating.''

After lunch, we leave Maranello and head for the mountains. Although I have not driven a Mondial for almost five years, the cockpit instantly fits me like a glove. Tall people will welcome the more generous adjustment range of the body-hugging seats and the three-spoked steering wheel. They will also praise the much-better-spaced pedals, which are now more or less in front of the torso instead of offset to the right. The new GTO-style instruments are a lot nicer than the previous design, but they do reflect badly in the windshield at night.

The 3.4-liter engine is a gem. It obeys throttle inputs with the vigor of a Doberman, it eagerly spins to its 7500-rpm cutout speed, and it has lots of top-end punch. Unlike most high-performance units, the 32-valve V-8 from Maranello remains smooth, docile, and relatively quiet throughout the entire rev range. Its predecessor ran out of steam at 4500 rpm, where it started to sound harsh and stressed, but the 3.4-liter version delivers fluid, turbinelike power all the way to within 1000 rpm of the redline. Yet despite this irresistible high-speed urge, there is also plenty of low-end torque on tap. Pulling away from 1500 rpm in top gear is a totally fuss-free exercise, and even in stop-and-go traffic, the Mondial t won't hiccup or sputter as so many oversexed exotics do.

On a dry, flat, straight, and preferably empty stretch of tarmac, the 3439-pound coupe can accelerate from 0 to 60 mph in 6.4 seconds and eventually reach a maximum speed of 158 mph. The average fuel consumption works out to 16.4 mpg, but leadfooters are unlikely to squeeze more than 300 miles out of the 25.4-gallon tank.

The clutch is perfectly balanced and not all that heavy. It feels meaty and powerful enough to tempt some first-gear smoke out of the fat Goodyear Eagles, an exercise that in Italy still draws applause rather than the attention of the police. Full marks again for the four disc brakes, which are sharp, strong, and progressive. Praise also goes to the power steering; it's effortless but still precise and communicative. The new five-speed gearbox, on the other hand, is as balky and stubborn as ever. I love the chromed gate and the traditional spindly lever, but I hate the slow and uncooperative shift action. The adjustable dampers don't work well enough, either. Although the hard setting provides a nice blend of poise and ride comfort, the test car repeatedly bottomed out at soft and medium settings.

Pushing the Mondial t through its paces is much more exciting—yes, even intoxicating—than setting spurs to a Porsche Carrera 4 or a Corvette ZR-1. Whereas the Porsche and the Corvette have been honed and tamed to perfection, the old-school Ferrari is sharper, more demanding, and less forgiving. It turns in with the determination of a switchblade knife, it holds the road well despite having less advanced limits of adhesion, and it handles the way mid-engined cars do. No dethrottle maneuvers in the middle of a ten-tenths bend, please, and no silly weight transfers, or the tail will come unstuck and swing around like a screeching pendulum. This adrenaline-producing attitude may not be to everyone's taste and skill, but serious drivers will find it entertaining and rewarding.

The old Mondial was much better than its reputation, and the new Mondial t is better still. To me, it is the most desirable Ferrari this side of an F40, and for about $82,000, it certainly is one of the few remaining buys in the exoticar league. If you are ready for the challenge, grab one now before the word gets around and the line grows long.

The 32-valve V-8 from Maranello remains smooth, docile, and relatively quiet throughout the entire rev range.

Testarossa looks and performance for new 348

It's taken Ferrari 14 years to get round to replacing the 308/328 series — but the new car is well worth the wait. Werner Schruf was the first behind the wheel

WHEN IT COMES TO TOUGH acts to follow, not many are tougher than Ferrari's 328, that gem of a supercar with gorgeous looks, a wondrous V8 engine and the sort of handling that makes you long to leap out of bed and go for a long Sunday morning drive.

But its successor — the 348t — is now here, unveiled to universal acclaim at the Frankfurt Motor Show. Its looks alone — the work of Pininfarina, once again — are enough to make the queues at Ferrari dealers even longer. And a short drive at Ferrari's Fiorano test track, before the show cars trekked off to Frankfurt, revealed that the 348t does even more for the spirits than the 328 when you slip into the leather-packed cabin and fire up the V8.

While the 348's V8 engine is mounted longitudinally (instead of transversely in the 328), the transverse gearbox is right behind it at the same level as the crankshaft and actually underneath the driveshafts. This arrangement, also used in the current Ferrari Formula 1 cars, means the 348's V8 sits more than 5ins lower than in the 328. And there, at a stroke, is a major step forward in Ferrari's quest to lift cornering performance.

The strakes along the doors, adopted from big-sister Testarossa, highlight another significant innovation in this eight-cylinder Ferrari series. The radiators now occupy the space liberated by fitting the engine longitudinally. There are water coolers on each side and an oil cooler on the right. The aggressive-

348tb shows Testarossa influence in side strakes but points to Ferrari's future in front-end styling. Interior is roomier than the 328's

looking new grille in the front is essentially only for decoration.

Like all current Ferraris, the 348 carries the unmistakable stamp of the Turin design house, Pininfarina. This new body is already being hailed as a masterpiece. The edges and corners that characterised the previous model have given way to soft curves. The front, with its distinctive grille and the large spot lamps integrated into plastic bumpers remind you a little of the F40 and will apparently reappear as typical features in future Ferrari models. The new body, made of steel apart from the aluminium bonnet and engine cover, has a beefy rear with black ribs over the broad rear lights and powerful curved mudguards over the wide rear wheels.

The exterior mirrors, the engine cover strewn with ventilation slits and the rear windows which curve inwards are reminiscent of the 288 GTO. The windows slope into two openings

through which the engine draws its air. As with the Testarossa and the GTO, Pininfarina has again succeeded in developing a harmonious shape without large spoilers.

Aerodynamically, the 348 does very well: the flush windows, the small air duct on the right side of the front spoiler for the air conditioning radiator and the largely flat underbody all contribute to the drag factor of 0.32.

Under the elegant skin, there is almost revolutionary innovation in the body construction. The old tube frame idea has been largely flung out. There is now only an auxiliary frame to support the drive and rear axle. The rest of the body rests on a self-supporting frame which is stiffer and can be made more accurately. "The 348's torsional strength is superior to that of all our other models," says Fabio Giunchi, Ferrari's design chief. "We concentrated development on the spyder version so it would be equally satisfactory in the open model as in the fixed-head. Of course, the spyder isn't quite so stiff as the berlinetta."

There's good news for tall Ferrari drivers: the 348 is comfortable. Its interior is 4ins longer and slightly higher than the 328's. The seats, which do not have a particularly good shape, can be pushed back far enough to allow decent room for long legs. An adjustable steering wheel is standard and can be raked at a steep angle, although it then slightly obscures the revcounter and speedometer.

The newly designed cockpit is very neat. The oil pressure and water temperature gauges are framed by the tachometer and speedo, directly ahead of the small leather steering wheel. The fuel and oil temperature gauges are in the central console and the dashboard extends to the doors and is covered in leather.

When you start driving, the clutch needs a good push and the gear stick is as stuborn as a mule until the gearbox oil has warmed up. The new transmission doesn't alter the typical Ferrari gearshift feel.

"Our sports cars should give the driver the feeling that they want to be mastered by him," says Ferrari president Piero Fusaro. So you have to give it a bit more than some of its rivals. The steering is no exception: Ferrari made a conscious decision not to include servo assistance, so powerful arm muscles are useful. Speed makes the work a bit easier. Then, its directness and precision makes very accurate driving possible.

The rest comes from the low centre of gravity, the stiffer pivot point for the trapezium and triangular suspension arms, high-performance Bilstein shock absorbers and, of course, the opulent tyres (front 215/50 rear 255/50) on 7.5 and 9ins wide, 17ins diameter wheels. The result gains your immediate respect: the 348 attains enormously high cornering speeds with hardly any roll, reacts quickly to changes of direction and has excellent traction, at least on a dry surface. Finding its ultimate potential is going to be hard on the open road.

It's also clear that 348 buyers shouldn't be too keen on comfort. The very stiff chassis tuning dispels any doubt: the new Ferrari is intended to be a high-class sports car without compromise. The sound of the V8 confirms this. It is considerably quieter than its predecessor, especially at low revs, because of substantial noise reduction measures and catalytic converters, but the characteristic sound is still there. From about 4000rpm, that metallic, aggressive fortissimo sets in; music to Ferrari-lovers' ears.

The 32-valve 3.4-litre V8 has astonishing qualities. Its usable rev range stretches from 1500 to 7500rpm, but it responds instantly to the accelerator. So despite the car's not inconsiderable weight — 3071lb — superior performance is assured. Ferrari claims a 0-62mph time of 5.6secs and a top speed of 171mph.

The 348 will make its UK debut at Earl's Court's Motorfair next month with sales likely to start at the end of the year. A right-hand drive model has already been type approved. There's been no official confirmation of UK prices but Italian sources suggest close to £55,000 for the 348tb and £57,000 for the 348ts.

FERRARI 348

SPECIFICATION
ENGINE Longitudinal mid engine, rear-wheel drive.
Capacity 3405cc, 8 cylinders in vee.
Bore 85mm, **stroke** 75mm.
Compression ratio 10.4 to 1.
Valve gear dohc per bank, 4 valves per cylinder.
Fuel and injection Bosch-Motronic ignition and fuel injection.
Max power 300bhp (PS Din) (221kW ISO) at 7200rpm.
Max torque 238lb ft (323Nm) at 4200rpm.

GEARBOX
Five-speed manual.

SUSPENSION
Front, double unequal length wishbones, coil spring, anti-roll bar.
Rear, double unequal length wishbones, coil springs, anti-roll bar.

STEERING
Rack and pinion.

BRAKES
Front ventilated discs. **Rear** ventilated discs. Antilock standard.

WHEELS AND TYRES
Front, 215/50ZR17 tyres. **Rear**, 255/50ZR17.

DIMENSIONS

Length	931ins (4230mm)
Width	417ins (1894mm)
Height	258ins (1170mm)
Wheelbase	540ins (2450mm)
Weight	3071ins (1393mm)

PERFORMANCE (claimed)

0-62.5mph	5.6secs.
Maximum speed	+171mph

FUEL CONSUMPTION
36.7 constant 56mph. 28.5 constant 75mph.

PRICE
Total (in UK) £N/A

FERRARI 328GTS

Ferrari's best-selling car ever prepares for retirement.

BY JOHN PHILLIPS III

PHOTOGRAPHY BY COLIN CURWOOD

Elkhart Lake, Wisconsin—

Ferrari 328 bashing is fashionable these days, especially in concupiscent communities like Newport Beach and Palm Springs, where the cars are more often spotted in the drive-through window of an In and Out Burger than they are carving corners on Laguna Canyon Road. In those towns, people who have recently had their hair fashioned into greasy purple spikes incite more interest than do drivers of Ferrari 328s. Even Coco Chinetti, son of Luigi, Sr., recently referred to 308s and 328s as "modern Ferraris, which are all Fiats."

All of which is unfair. With each model year since the original 308's introduction at the 1975 Paris Motor Show, Ferrari's engineers have improved the car steadily. Nowadays, the 328GTS and GTB are so refined and reliable that—and this somehow seems blasphemous—the car works almost effortlessly as a day-to-day commuter.

And yet . . .

And yet there remains that edge. In a 328, you don't feel all cozy and comfy as you might in, say, a Porsche 928S4 ($74,545) or even a 911 Turbo ($70,975), two of the 328's closest competitors. Part of the Maranello formula is a kind of shameless machismo built into every mechanical function: The Ferrari's brake pedal is heavy. You won't want to hold the clutch depressed through even one red light. Under 30 mph, the steering requires so much muscle that the whole rack groans and shifts uneasily in its brace. And the infamous metal-gated shifter refuses to cooperate at all unless you've performed a perfect blip-and-slip ballet in proper sequence. Like so many of the 328's controls, the shift lever must be deliberately *placed,* not pushed. Engine and transaxle revolutions must agree. On cool mornings, the 328 wants to idle for five or ten minutes before *any* gear may be selected. In no car but a Ferrari (and maybe a $67,500 Lotus Esprit Turbo) are you made so persistently mindful and respectful of the shiny internal bits.

In fact, Ferrari more or less *ensures* you'll pay attention. There is no optional cruise control, no automatic transmission, no adjustable steering wheel. The seats are hard, narrow, and flat, and are adjustable only for rake. The steering feels as if it had been lifted intact from a

Formula Ford: fast, direct, with plenty of kickback and an awkward 40-foot turning circle. Move the wheel an inch off center, and you're in the ditch. To be certain you get the point, the factory doesn't even deliver the car with a radio.

Of course, during a seven-day, 2850-mile trip in the 328GTS, we never really missed the music. There is, instead, the music of the engine, which tends to be the principal and dominant performer anyway. It emits not so much a mechanical growl as a constant throb, which becomes particularly insistent—damnably *resonant,* to be blunt—from 3500 to 3800 rpm. That's unfortunately the 75-mph region, a common fifth-gear cruising speed. To lessen cockpit boom, the driver involuntarily lets the engine creep up to a happier 4000 rpm, or 90 mph.

The message is clear: "You want to pretend you're Gerhard Berger, pal, then *pay attention.*"

And at that velocity, the 328, at least on America's Interstates, becomes conspicuous radar prey.

If there's a more tractable V-8 around, we'd like to hear about it. Go ahead and shift from first to third—the 328 doesn't care. From a dead stop, you can drop the clutch at 2000 rpm with the selector in second and then hold your foot to the floor for another 5800 rpm. Power is transmitted in a smooth, uninterrupted arc that begins as the aforementioned throb and then turns into a whoop, a wail, and finally a ripping scream. Throughout the range, there are no flat spots or noticeable surges.

For the first hour in the car, you hold each gear to 6000 or so rpm just to enjoy the racket. Later, when you discover that there's plenty of power available at far fewer revs, you find yourself short-shifting. Zero to 60 mph is a 6.6-second piece of business, depending entirely on your skill at negotiating the tricky first-to-second shift. (First gear is on the lower left, outside the conventional H.) Overtaking is so effortless, even at 60 or 70 mph on country roads, that you tend to leave the selector in fifth. A downshift only encourages onlookers to jot down your license number.

"When Americans think of great V-8s," noted one 328 owner whom we met in Wisconsin, "they usually think of the Corvette's as the most willing, as the beefiest. And yet the Corvette has only a 3000-rpm power band—essentially from 2000 to 5000 rpm—which makes it feel practically peaky and temperamental compared with the 328's." He's right. Which kind of makes your heart race at the thought of all the small-displacement four-valve V-8s on the way from Japan.

Until insane speeds are reached, the 328's handling is dead neutral, with the near-perfect balance of a Porsche 944S. If pitching a car is your thing, then go ahead; the suspension will tolerate it. But the 328 doesn't want to be thrown, chucked, or hurled through corners. It rewards smoothness. Trail-brake gently into turns; then squeeze on the throttle. The racing-car-style steering telegraphs through your fingertips every pebble, every change of tarmac, every subtle nuance of road surface. Shrieking through twisty Kettle Moraine State Forest, southwest of Plymouth, Wisconsin, we can recall no turns that couldn't be negotiated comfortably at 70 mph. While driving the 328, you continually have a clear sense of what each wheel is doing—how much weight it's bearing, how much more braking or steering or power it can gracefully accommodate. You begin to feel invincible.

Resist that feeling.

Attain truly frightening speeds—at least on dry pavement—and the 328's nose begins to plow. Get back on the power smoothly, delicately, and your passenger will think you're Phil Hill. Get back on the power too hard and too soon—especially on rough or broken pavement—and the understeer is displaced in a flash by World of Outlaws oversteer. It's another clear message from Maranello: "You want to pretend you're Gerhard Berger, pal, then *pay attention.*" Our advice is that you become acquainted with this schizophrenia on a rainy Sunday afternoon in the empty parking lot of the domed stadium in your area.

The ride is rigid, much like that in the stiff-legged Porsche 911. But the high-frequency jitters—the small-displacement road harshness that becomes so annoying at the end of a long haul in a 911—have been largely filtered out of the Ferrari. Five hundred miles a day in the 328 is no problem.

There are, in fact, a *lot* of pleasant surprises here. The stowage compartment behind the engine holds enough luggage for two people for a three- or four-day trip. It gets toasty back there, however, so carrying perishables like Upper Peninsula peaches is a rotten notion. Removing and installing the lift-off roof is a mindlessly simple forty-five-second snap-and-stow job. Topless, the 328 remains more solid, more of a piece than any opened-up coupe we've driven save the Porsche 911 Targa. No shake, rattle, roll, or twist.

Visibility fore and aft is surprisingly good. The backlight allows a small but unobstructed view astern. You actually don't need the side-view mirrors to back into a tight parking stall. But we still wish the garish louvers on the quarter-windows were removed, if only to gain more side visibility.

The air conditioner, even on a day when the temperature topped 100 in Madison, Wisconsin, chills the cockpit like the cooler you'd find in a Cadillac Brougham. The engine started first time, every time, even after we par-boiled it for thirty minutes in a stifling traffic jam at the northern end of the Mackinac Bridge. Not a single piece of interior trim came adrift or failed to operate as advertised during the car's 2850 miles with us. Indeed, the fit and finish, for the first time we can recall in any Ferrari, were beyond reproach. All of this in a car whose entire 12,000-mile history had been spent in the usually uncaring hands of journalists and test drivers. As far as we know, only Honda and Toyota allow their press fleet cars to remain "in rotation" that long.

Given our remarkably positive experience with the 328, it is thus with fondness (and more than a little sadness) that we prepare to bid it *arrivederci*. Nineteen eighty-nine marks the 328's final model year, a year in which anti-lock brakes make their first appearance on a Ferrari of any kind and a year in which the 328's price—inevitably, inexorably—climbs another $5900. After that, Ferrari's best-selling car will be replaced by the 348, a vehicle that spy photos depict as a seven-eighths-scale Testarossa. Indeed, powered by a 300-bhp, 3.4-liter V-8, the 348, according to Ferrari North America's Dr. Emilio Anchisi, "will feature performance that *exceeds* that of a Testarossa."

Well, maybe it will. And, sure, we agree that the 328 is in need of a styling update. But in the interim, we're perfectly content to enjoy a car that Ferrari has pretty well perfected in the preceding fourteen years. The 328 is a sports car that is finally as fast and fun as it is reliable and—should we even *think* it?—rational.

	POOR	FAIR	GOOD	EXCELLENT
ENGINE				
power				•
response				•
smoothness				•
DRIVETRAIN				
shift action		•		
power delivery				•
STEERING				
effort		•		
response				•
feel				•
RIDE				
general comfort		•		
roll control				•
pitch control				•
HANDLING				
directional stability		•		
predictability				•
maneuverability			•	
BRAKES				
response				•
modulation		•		
effectiveness				•
GENERAL				
ergonomics		•		
instrumentation				•
roominess		•		
seating comfort			•	
fit and finish				•
storage space	•			
OVERALL				
dollar value		•		
fun to drive				•

FERRARI 328GTS

GENERAL:
Mid-engine, rear-wheel-drive coupe
2-passenger, 2-door steel body
Base price/price as tested $77,900/$77,900

MAJOR EQUIPMENT:
Air conditioning standard
Sunroof (lift-off targa panel) standard
Leather interior standard
AM/FM/cassette available as extra-cost dealer-installed item only
Cruise control not available

ENGINE:
32-valve DOHC V-8, aluminum block and heads
Bore x stroke 3.27 x 2.90 in (83.0 x 73.6mm)
Displacement 194 cu in (3185cc)
Compression ratio 9.2:1
Fuel system Bosch K-Jetronic mechanical fuel injection
Power SAE net 260 bhp @ 7000 rpm
Torque SAE net 213 lb-ft @ 5500 rpm
Redline 7700 rpm

DRIVETRAIN:
5-speed manual transmission
Gear ratios (I) 3.39 (II) 2.35 (III) 1.69 (IV) 1.24 (V) 0.92
Final-drive ratio 3.82:1

MEASUREMENTS:
Wheelbase 92.5 in
Track front/rear 58.1/57.7 in
Length 169.1 in
Width 67.7 in
Height 44.1 in
Curb weight 3170 lb
Weight distribution front/rear 44/56%
Fuel capacity 18.5 gal

SUSPENSION:
Independent front, with unequal-length A-arms, coil springs, anti-roll bar
Independent rear, with unequal-length A-arms, coil springs, anti-roll bar

STEERING:
Rack-and-pinion

BRAKES:
10.7-in vented discs front
10.9-in vented discs rear
Anti-lock system

WHEELS and TIRES:
16 x 7.0-in front, 16 x 8.0-in rear cast aluminum wheels
205/55VR-16 front, 225/50VR-16 rear Goodyear Eagle VR tires

PERFORMANCE (manufacturer's data):
0–60 mph in 6.6 sec
Standing ¼-mile in 14.6 sec
Top speed 155 mph
EPA city driving 13 mpg

MAINTENANCE:
Headlamp unit $76.00
Front quarter-panel $1986.68 (available as entire front clip only)
Brake pads front wheels $144.90
Air filter $37.42
Oil filter $27.63
Recommended oil change interval 7500 miles

By PETER ROBINSON

Ferrari, nein danke

Dear Santa,

I've always thought you were a lovely bloke, generous to a fault, and I'd hate to see you make a mistake. Yes, I know I've always craved a Ferrari and anticipated seeing one in my stocking on Christmas morning. But not this Christmas, Santa, not if it's going to be one of the new 348s.

So, just in case you've managed the impossible, and propose slipping one into my garage on Chrissy eve, I'd just as rather you didn't, thanks all the same. This year you give 'the Fazz' to some other, less finicky, person.

I'm sure you're as surprised as I am at my late, much agonised over, change of heart. It must be the first year for as long as you can remember that I haven't asked for a Ferrari.

Saying *no* to a 348, when everything points to this Christmas being *The One*, is probably going to be hard for you to understand. After all, I do live in Italy now and the factory's only 90 minutes down the Autostrada and I know how easy it must be for you to organise a 348 for a truly deserving chap such as myself.

You'll want to know why, of course. That's not so easy to explain and may even be somewhat illogical, because there is so much about the 348 that is exciting and truly wonderful. Maybe I'm simply getting older.

Santa, I love the look of the car. It couldn't be anything but a Ferrari. Pininfarina has been able to lower the cd to 0.32 without any add-on spoilers, so the lines are pure and clean, yet compelling and quite beautiful from some angles. The links with the TR – that's a Ferraristi's short hand for Testarossa – are all too obvious. And that's what makes it predictable, even conservative. Sometimes I wish Ferrari would commission other Italian design houses to come up with a new and stirring interpretation of how a Ferrari should look. Even before I saw the first scoop photographs of the 348 I kind of knew what to expect, but it lacks the creativity, even the voluptuous curves, that I reckon every Ferrari should have.

It's immediately obvious the 348 is wider that the 328, with far less rear overhang. In fact it's slightly shorter than the old car, yet 5.3ins broader and the wheelbase has grown by 4ins to exactly 100ins. A TR is just 3ins wider. The reasons for the added width are

Saying no to a 348 is probably going to be hard for you to understand

not hard to find, for the Ferrari engineers have moved the radiators from the nose to beside the engine, which also helps explain those huge and rather ugly air intakes in the doors and the cleverly disguised vents just behind the trailing edge of the small rear side windows. So while the 348 does have a grille, it's a phoney, even if Ferrari says it invokes memories of those classic Ferraris like the 375MM and 250P.

You will have guessed that the 348 follows the Mondial in having a new 3.4-litre dohc V8 engine, that is mounted longitudinally with the gearbox at 90° to the engine. This positioning of the engine means it has been possible to lower it 5.1ins compared to the transverse engine in the 328 and this helps lower the centre of gravity and improves accessibility.

Official Ferrari figures suggest the 348 has the same 300bhp at 7200rpm as the Mondial. Unofficially, the Maranello engineers concede that it actually develops a little more grunt that the two plus two model. Of course it does have a 280lb weight advantage over the Mondial, but even so the 348 is rather heavier than I expected. The early 'dirty' versions weigh 3065lbs, an increase of 282lbs over the 328GTB. Whatever happened to the philosophy behind the fibreglass 308s, that tipped the scales at 2530lbs?

A wonderful engine

I adore the engine. Strong and always energetic, the 32-valve, small capacity V8 accepts 1000rpm in fifth gear, yet thrives on revs. Below 5000rpm there is progressive build up in gear whine and an exhaust note that emphasises

Far left and left: The 348 is wider than the 328 and Pininfarina has been able to lower the Cd to 0.32. Below left: The 32-valve V8 accepts 1000rpm in fifth gear. Below centre: 348s for the Christmas stocking. Below right: Official heritage courtesy of Ferrari. Bottom: Internally, the 348 is pure sports car.

merely the wonderful smoothness of the engine. Spirited it may be, but the volume is still acceptable on an Inter-city journey.

At five grand the engine's song is transformed, filling the cabin with a cacophony of cam wail and exhaust scream. Justifiable over a two hour stretch, when this symphony of sound will be music to the ears of Ferrari purists, the level of noise impedes the 348's competence as a serious high speed touring car since 5000rpm corresponds to 110mph.

Ferrari says the 348 is quicker around Fiorano than a five-litre, 390bhp TR, and I believe them. Likewise, the factory claims the 348 will accelerate to 62mph in 5.6s, 0.2 secs faster than its big brother. It feels at least that quick, helped no doubt by relatively low gearing that means the car pulls to the 7500rpm redline in fifth gear – 166mph – yet is so tractable that the 348 is remarkably easy to drive, despite the lack of power steering.

Ferrari says the people who buy 348s want a pure sports car, and is keen to infuse the all-important tactile responses. Not surprisingly, the steering is heavy at low speeds and not as sharp at the straight ahead as you might suppose. Still the 348 turns in flawlessly, the steering loading up progressively, constantly relaying information back to the driver.

Massive 255/45ZR17 rear and 215/50ZR17 front tyres provide phenomenal grip, though, even with a 40% limited-slip differential it's still possible to furiously spin both rear wheels. I've no arguments with the new Ferrari's handling. It's more progressive and more predictable than the 328, with just a brief moment of understeer when entering a corner, before it turns neutral. Only excessive power can induce oversteer. Even so, the skilled can adjust this with the right foot, giving the 348 a delicious nuance of sensitivity.

No Santa, my big argument with the new Ferrari concerns not the engine, nor the roominess of the cabin, the roadholding or even the styling, but the ride quality. Now, I know I will be accused of losing sight of the car's heritage and the basic principles under which it was created, but there simply is no reason why it needs to ride as firmly as it now does. The admittedly early build car I drove lacked the suppleness necessary to cope with undulations in the road surface at Fiorano, while the front end developed an unsettling edginess over ripples. Taut suspension and damper settings mean the car conveys exactly the condition of a normal road surface.

Santa, I know there are problems with any new car – Ferrari has admitted it is having difficulty getting the early 348s up to scratch – and just a couple of weeks ago there were about 35 undelivered 348s littering up the parking area around Maranello waiting for minor corrections to go through before they could be delivery to impatient customers around the world.

This morning I drove a new TR and the improvement over the original cars is remarkable. The ride is far more resilient, the quality now outstanding.

No Santa, if you don't mind I'd rather wait until Christmas 1990 before finding a 348 in my stocking.

Thanks for understanding.

Yours, Peter. ■

FERRARI 348 tb

CAR's testers are not immune to the Ferrari syndrome, the insidious intoxication that makes even hardened sceptics flip their lids, when exposed to the cars from Maranello. Stuart Johnston, fortified with quotes from colleagues Dave Pollock and John Wright, describes a one-day stand with the latest, sexiest coupé...

IT IS impossible to begin a test drive in a Ferrari with the mental composure you would reserve for, say, a Mazda or even a Mercedes-Benz. So many millions of words have been written about these missiles from Maranello; the countless superlatives, ancient tales of the Mille Miglia and Monza, grumpy old Enzo's idiosyncrasies, cam-whirrings, multi-cylinder howls and the seductive qualities of Ferrari sheetmetal all calculated to spin your head.

Having said that, the new 348 tb is so beautiful it should be illegal. It is almost certain that Enzo Ferrari would have clutched this car to his bosom but we'll never know as this is the first all-new model to be produced by the Ferrari factory since the Great Man's death in 1988. We feel it is one of the most classic sports car designs to emerge from the 1980s, a composition of artistry and function, speed and aggressiveness; and loaded to its air ducts with sex appeal.

The 348 tb that CAR sampled recently arrived painted red (surprise), with massive silver coloured alloy five-spoke rims, Bridgestone RE71s of 225 mm footprint and an aspect ratio of just 45 per cent and Cape Town Ferrari agent Luigi Viglietti in the passenger seat. As this 348 was a pre-production model, one of just a handful in the world, there were strict instructions from the very peak of the Ferrari empire that no journalist should be let loose on his own in the car.

PICTURES BY STEVE TRONSON

You don't just hop into a Ferrari and roar away when one is offered for road impressions. There is a compulsion to poke and pry at the car's innards, to step back and admire it from all angles, to discuss the styling, performance potential and engineering of this new mid-line model.

Stretched V8

Raising the hood reveals the now-longitudinally-mounted, dry-sump V8 which has been stretched to 3 405 cm³. The 32-valve engine delivers 225 kW at 7 000 and 323 N.m of twist at 4 200 revs. The standard of workmanship is impressive and the purist will fall in love with the layout of air cleaner ducting, wiring, the crackle

The prototype was shod with optional Bridgestone RE71s and production models feature Ferrari's own brake calipers (top). In profile (above) the 348 resembles a Testarossa. The fluting on the sides directs air to twin side-mounted radiators.

finish on the cam covers, the flowing exhaust header tubing and the integration of the rear chassis sub frame and engine, which according to Viglietti, makes working on the motor a piece of tapioca.

The biggest mechanical change with the 348 is the mounting of the motor longitudinally in the car, whereas the old 308 and 328 had transverse V8s of similar design. This was achieved by fitting a transverse five-speed gearbox behind, rather than beneath the motor. Thus the added length of the longitudinally-mounted motor is minimised by the "short" box and the motor can be slung low down in the chassis tubes to give an optimum centre of gravity. Again, access to serviceable components is far better than it was with the 308 and 328 "transverse" V8 models, particularly for plug changes. Incidentally, the "tb" suffix on the 348 refers to the car's transverse gearbox and stands for "transversale berlinetta".

A very advanced Bosch Motronic engine management system provides even low-speed fuel distribution and precise ignition timing. There is a separate hot wire sensor and ignition system for each bank of cylinders for optimum control of combustion and the 32-valve cylinder heads provide excellent flame distribution for easy city running and hard-on high speed pursuits.

Finely balanced suspension

Pressed steel is now used in place of forged steel units for the double wishbone suspension, which provides anti-dive geometry at the front and anti-squat at the rear. It is interesting to note that no electronic damping control is provided for the suspension, although a lot of attention was paid in the design and testing stages to obtain a fine balance between ride comfort and stability.

The prototype model we sampled was fitted with the Teves-developed ABS braking system which is an option on production cars. The ventilated discs use double-piston calipers made from aluminium alloy and contribute to keeping unsprung mass down, as do the new lightweight suspension components.

Sheetmetal is used for most of the body and chassis unit, apart from the drive train subframe. However, the fluting on the sides used for ducting air to the twin side-mounted radiators is of a composite material and the bonnet and luggage compartment hood are made from aluminium, to

‹eep overall mass (dry) down to just 1 393 kg.

One of the strangest aspects of the 348 tb is that it comes without a spare wheel. According to our Italian correspondent, Giancarlo Perini, Ferrari has discovered that owners "do not stop to replace a flat wheel any longer" and the decision was made to provide extra luggage space in the nose, as well as to shorten the rear overhang behind the motor. Weird.

On the road

Being our resident Ferrari freak, Features Editor Dave Pollock was told "Don't let your enthusiasm bias your impressions." But Dave points out that a large part of the measure of a car like the 348 is your emotional reaction to it and from a visual point of view, all of CAR's staff agreed that this was one of

The bull terrier-like stance of the 348 is achieved through a massive 165 mm increase in width over the 328's 1 729 millimetres, giving the new mid-line Ferrari an air of aggro. The new longitudinal location of the 225 kW, 90-degree V8 (left) has been achieved with a transverse gearbox.

We were knocked out by the simplicity and function of the Ferrari cockpit (above, left) where the yellow Ferrari steering wheel badge draws the attention. But that tall gearshift is recalcitrant. Opening the doors reveals Kevlar radiator ducts (above, right).

the most stunning, if not THE most stunning, motor cars ever to park in our Pinelands yard.

"Grab the underside of the door flute," says Dave, "press the tiny door button with your thumb and the intimate, functional cockpit is revealed. The seats are minimal yet superbly supportive and the planes of the interior surfaces interface with a confident simplicity born of the true artist's eye."

The simple three-spoke steering wheel manages to be both classic and modern in its design and is just the right size, with a medium width rim clad in firm black leather and a yellow Ferrari logo in the centre. A mixture of leather and vinyl is used for trimming the all-black cockpit and apart from a rather shoddy console pocket-cum elbow rest, the standard of workmanship was very high on this pre-production model.

The large analog dials that are visible behind the wheel rim read up to 320 km/h in the case of the speedo and 10 000 r/min for the tachometer, with red-line beginning at 7 500 revs. Oil pressure and water temp gauges are contained in the instrument pod, while to your right on the console – this being a left-hand-drive car – are oil temperature and fuel gauges. The switchgear for lights and flickers is a bit flimsy-looking, much like that fitted to the Lancia Thema, and the stalk light switch is a bit cumbersome.

There is central locking fitted as well as air-conditioning and electric window winding and the luxurious approach is far removed from the cable door handles fitted to the stark, function-first F40. Clearly this is meant to be an everyday Ferrari.

'Disconcerting position'

"The driving position," said the analytical and solidly-built John Wright (CAR's Technical Editor) "is initially disconcerting because it is so different.

Ferrari country: an empty road, majestic scenery and time ...

I pushed the seat back but my knees were still on either side of the steering wheel. Luigi told me I could adjust the steering column for extra height but I felt that making the wheel higher would be less desirable."

Smaller people would not have too much trouble in this regard, although the footbox requires the wearing of narrow shoes as the pedals are placed very close together and at a steep angle, which takes some getting used to.

Starting the motor involves no pumping of the accelerator pump as it used to in the days of gaping Weber juice gobblers. Keep your foot off the throttle, twist the key and the 90-degree V8 comes alive with surprisingly little drama. It idles smoothly and sounds unlike an American V8 with no hollow burbling but rather a crisp, metallic rasp that is not unlike a highly tweaked four-cylinder.

If the motor's sound was a surprise, the first major shock comes when trying to select first gear, which is towards you and to the rear of the bus-like gate that Ferrari has favoured for all these years. The gear linkage, despite being totally new for the transverse gearbox layout, is a throwback. Amazingly, Ferrari has managed to build in all the stiffness and recalcitrance that has been "challenging" the owners of its cars for decades.

"The clutch takes rather high" comments Dave Pollock and this is something that all three of us agreed on, with John Wright complaining that his accelerator foot snagged on a piece of carpeting, making him uneasy at times. There is no need to slip the clutch on pull-off as the low down torque is prodigious.

Notchy gearshift

"Floor the throttle at 40 km/h in fourth gear and the 348 tb just surges forward," recalled Dave. "The heavy, notchy gearchange would give fans of

Japanese light cars an apoplexy and it's really a serious handicap at first."

It's definitely the most dominant feature of the car when you are acclimatising to it. A heavy amount of pre-load on the tall, spindly lever means that you cannot rush the changes, you have to let it find its own way across the gate and yet once you have found the right slot, a ridiculously hefty shove is needed to slot home into the next cog. All this tends to require too much concentration when you are threading a half-million rand car through rusty Third World wrecks and massive long distance trucks, but then Ferrari has never claimed that its cars are without quirks.

Once on the highway and with the lever safely in fifth, it's time to explore the speed potential. At parking speeds the steering is so light that one is misled into believing that there is some power assistance, but as speed rises the steering loads up and the light, low-speed steering is merely a function of clever geometry and very neutral mass distribution.

The big treat in the ride department is how well this firmly-sprung car on ultra-low profile tyres soaks up the bumps. The ride is almost in the limousine class and seems incongruous in relation to your proximity to the tar whizzing underneath the nose.

"I knew from the start that I would like the view of the road," said John. "It reminded me a little of the Toyota MR2 (sorry, Enzo), for the driver is low enough to see the undulations on the road surface well before the car contacts them. The screen is large and sloped and there is a panoramic view of the traffic."

Disappointing sound

Wind on just a touch of throttle and the speedo climbs to 180, then 200 with no hint of strain. The three testers were divided as to the quality of the engine sound. I was a bit disappointed, for I found that it stirred no particular feelings in me, reminding me more of a turbocharged four-cylinder than a complex piece of Italian engineering. John Wright said that the engine sounded aggressively metallic, although he could see that many would regard it as a "symphony of reciprocation."

Dave Pollock was ebullient: "A sustained plunge of the right foot releases a screaming torrent of sound from the 32-valve V8 which permeates every pore of your body and tickles the soles of your feet. But it's not quite as sweet as I had expected – certainly not a patch on the Testarossa's gutteral howl – and the revised valvetrain is perhaps too quiet now, robbing the car of some emotional Ferrari tradition."

The Ferrari 348 tb has a claimed top speed in the region of 270 km/h and we have little doubt that it will reach such a speed with no sweat or tears and with admirable directional stability. While traffic was always a problem during our single day spent with the car, we saw an indicated 250 km/h appear with contemptuous ease on the classically-styled speedometer, with speed still rising steadily at this point on a level road.

Accelerating to 100 km/h for our standard sprint runs would take a lot of familiarisation. The power is there for a low-to-mid-six-second run but the first-to-second gearshift would temper things dramatically, unless one was very skilled with the box, which we imagine would only come with lots of practice.

Steering loads up quickly

At 200 plus the steering has loaded up nicely and yet there is very little tramlining or bump steer. In fast sweeps, we found that there is lots of castor on the front steering geometry. The steering wheel loads up very quickly and you have to exert lots of pressure to maintain a line, when cornering hard. We all tended to like this trait, as it imparts lots of feel through the small leather rim and when traction limits are being approached, the steering will begin to go light.

However, the way this car has been set up, it is never likely to understeer. In fact you drive it on the throttle and it is quite easy to obtain a slightly tail-out cornering position, oversteer which is very easily corrected via the wheel or tempered by simply backing off the throttle. Putting more power on in a corner hangs the tail out just a touch more, backing off brings it in, and it is all so nice and predictable that a day in the mountains with this car should prove loads of fun and quite safe.

"Any change of direction is so quick and precise that it raises your perception of steering and handling response to a totally new plane," commented Dave. "The beauty of it is that the 348 feels incredibly alive and almost literally pivots around the driver." John Wright felt that the steering loaded up more than he liked in fast corners. "It then lightens quite a bit when weight is taken off the front wheels temporarily due to undulations, but it is dead easy to correct because the steering is very fast and reacts stably to rapid input.

"As neutral as it might be, there is an element of nervousness to its responses which can even be felt in wide sweeping corners; but this is surely how a fast Italian car is meant to feel. Of course, some impressions have to be tempered by the fact that one is travelling faster than one realises."

In twisty, hard charging the gear change is at its worst. This statement is perhaps a bit unfair because, as it is a left-hand drive car, South Africans are straight away at a bit of a disadvantage, although all three testers have driven quite extensively overseas in left-drivers. The biggest hang-up comes when you are charging from corner to corner, where an up-shift would be called for, followed by a down change under braking; with such a tricky box it takes too much concentration to risk going for the optimum gears and thus one's rhythm tends to be dictated by the gearlever mechanism rather than the road, the enormous reserves of handling or the power available.

After some time in the car one begins to learn how much pressure it takes to feel the gear in and indeed, when you get it right, there is an enormous sense of satisfaction. This is probably why Ferrari owners rarely complain about their gearshifts: they don't want to be thought incompetent.

We all enjoyed the brakes, which feature the Alfred Teves anti-lock system. Some hardboiled Ferrari owners apparently eschew the idea of electro-mechanical lock-up control but for everyday motoring we feel it is a necessity on a car of this potential. The pedal pressure required to slow the car is minimal and yet feel has not been sacrificed at all.

With so much glass to heat up the cockpit, air-conditioning is a must in this car for South African conditions, but there is a surprising air of roominess in the cabin, far more so than in comparable mid-engined cars from the likes of De Tomaso and Lamborghini. Ventilation is good but there is not too much space in which to keep odds and ends.

Summing up

In summing up, all three testers were impressed with the everyday nature of the car's behaviour. Good quality upholstery and interior design, allied to a very tractable powertrain, add up to a car that you could use for business, although it is a bit of a pain having to double up and stretch every time you want to climb in or out of the beast.

For me this Ferrari is at its best when viewed from the outside, as I find the styling so good that I don't think I could improve upon it. John Wright feels that it is somewhat educational to reflect that a car of this pedigree is no longer as far ahead of the fast end of the production sector as it would have been 10 years ago. And Dave Pollock says it is one of the most effective street driving machines he has ever laid hands on and undoubtedly the most thrilling.

One thing is for sure. Ferraris don't inspire half-baked reactions. And despite our criticisms, we aren't sure we'd want the 348 tb to be any different. For one thing, it wouldn't be half as much fun to write about. ●

MONDIAL t CABRIOLET

Vehicle type: mid-engine, rear-wheel-drive, 2+2-passenger, 2-door convertible

Price as tested: $97,000

Options on test car: base Ferrari Mondial t Cabriolet, $94,450; gas-guzzler tax, $1500; freight, $1050

Standard accessories: power steering, windows, and locks, A/C, tilt steering

Sound system: none

ENGINE

Type	V-8, aluminum block and heads
Bore x stroke	3.35 x 2.95 in, 85.0 x 75.0mm
Displacement	208 cu in, 3405cc
Compression ratio	10.4:1
Engine-control systems	2 Bosch Motronic M2.5 with port fuel injection
Emissions controls	3-way catalytic converter, feedback fuel-air-ratio control
Valve gear	chain- and belt-driven double overhead cams, 4 valves per cylinder
Power (SAE net)	300 bhp @ 7000 rpm
Torque (SAE net)	229 lb-ft @ 4000 rpm
Redline	7500 rpm

DRIVETRAIN

Transmission	5-speed
Transfer-gear ratio	1.14:1
Final-drive ratio	3.56:1, limited slip

Gear	Ratio	Mph/1000 rpm	Max. test speed
I	3.21	5.7	42 mph (7500 rpm)
II	2.11	8.6	65 mph (7500 rpm)
III	1.46	12.5	93 mph (7500 rpm)
IV	1.09	16.7	125 mph (7500 rpm)
V	0.86	21.2	159 mph (7500 rpm)

DIMENSIONS AND CAPACITIES

Wheelbase	104.3 in
Track, F/R	59.9/61.4 in
Length	178.5 in
Width	71.3 in
Height	48.6 in
Curb weight	3540 lb
Weight distribution, F/R	42.4/57.6%
Fuel capacity	22.5 gal
Oil capacity	11.6 qt
Water capacity	21.1 qt

CHASSIS/BODY

Type	full-length frame bolted to body
Body material	welded steel stampings and aluminum stampings

INTERIOR

SAE volume, front seat	49 cu ft
rear seat	28 cu ft
luggage space	4 cu ft
Front seats	bucket
Seat adjustments	fore and aft, seatback angle
General comfort	poor fair **good** excellent
Fore-and-aft support	poor fair **good** excellent
Lateral support	poor **fair** good excellent

SUSPENSION

F: ind, unequal-length control arms, coil springs, 3-position cockpit-adjustable electronically controlled shock absorbers, anti-roll bar

R: ind, unequal-length control arms, coil springs, 3-position cockpit-adjustable electronically controlled shock absorbers, anti-roll bar

STEERING

Type	rack-and-pinion, power-assisted
Turns lock-to-lock	3.0
Turning circle curb-to-curb	38.9 ft

BRAKES

F:	11.1 x 0.9-in vented disc
R:	11.7 x 0.8-in vented disc
Power assist	hydraulic with anti-lock control

WHEELS AND TIRES

Wheel size	F: 7.0 x 16 in, R: 8.0 x 16 in
Wheel type	cast aluminum
Tires	Goodyear Eagle ZR55; F: 205/55ZR-16, R: 225/55ZR-16
Test inflation pressures, F/R	38/39 psi

CAR AND DRIVER TEST RESULTS

ACCELERATION — Seconds

Zero to 30 mph	2.2
40 mph	3.2
50 mph	4.8
60 mph	6.2
70 mph	8.1
80 mph	10.2
90 mph	12.4
100 mph	16.1
110 mph	19.8
120 mph	24.0
130 mph	32.5
Top-gear passing time, 30–50 mph	9.0
50–70 mph	9.1
Standing ¼-mile	14.5 sec @ 98 mph
Top speed	159 mph

BRAKING

70–0 mph @ impending lockup	176 ft
Fade	**none** moderate heavy

HANDLING

Roadholding, 300-ft-dia skidpad	0.85 g
Understeer	**minimal** moderate excessive

COAST-DOWN MEASUREMENTS

Road horsepower @ 30 mph	7 hp
50 mph	17 hp
70 mph	35 hp

FUEL ECONOMY

EPA city driving	**12 mpg**
EPA highway driving	**17 mpg**
C/D observed fuel economy	**16 mpg**

INTERIOR SOUND LEVEL

Idle	65 dBA
Full-throttle acceleration	90 dBA
70-mph cruising	79 dBA
70-mph coasting	78 dBA

CURRENT BASE PRICE dollars x 1000

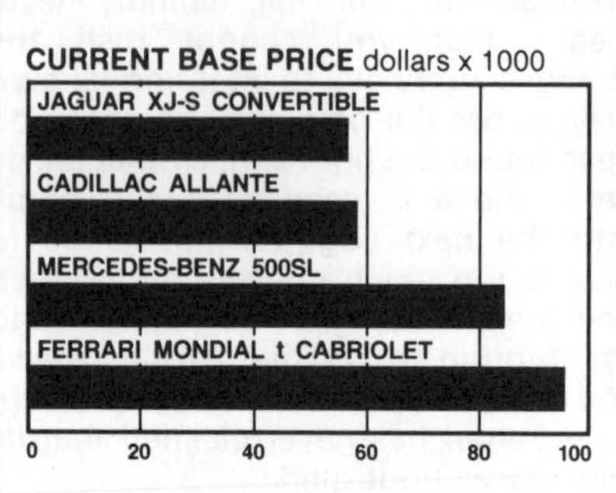

ACCELERATION seconds ■ 0–60 mph ■ ¼-mile

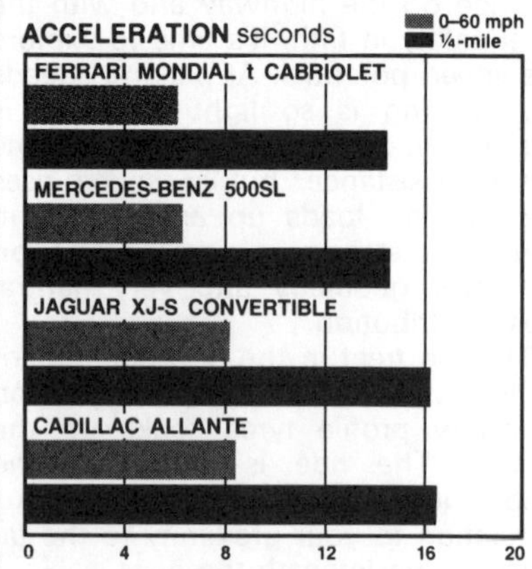

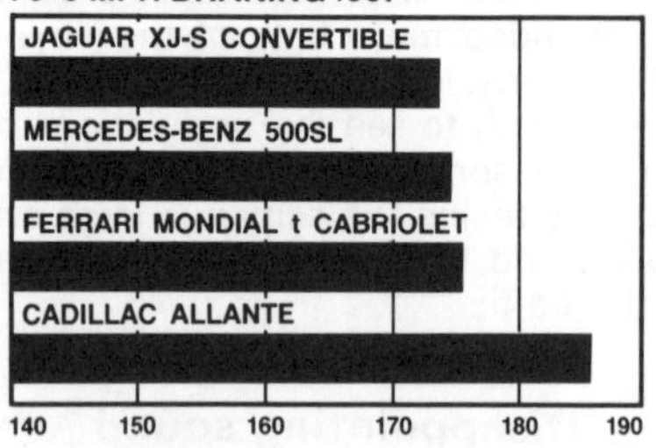

ROADHOLDING 300-foot skidpad, g

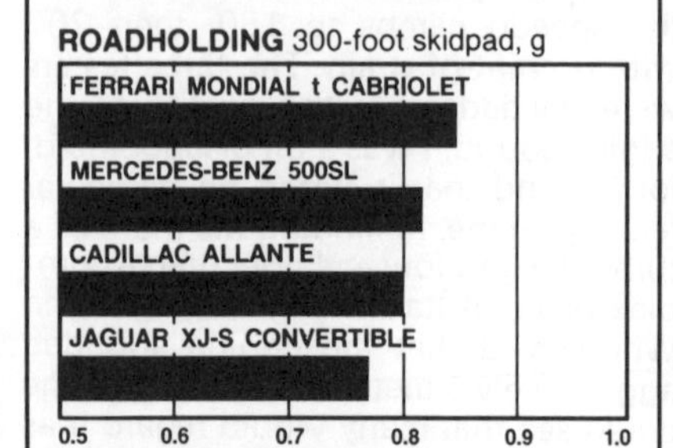

EPA ESTIMATED FUEL ECONOMY mpg

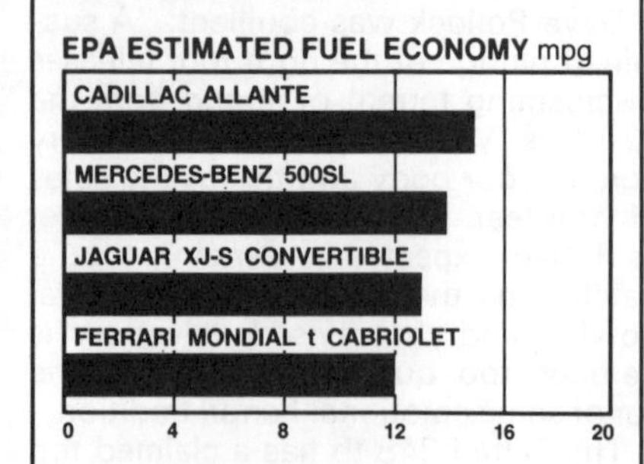

Ferrari Mondial t Cabriolet

The Formula 1 convertible.

BY CSABA CSERE

• Who among us hasn't dreamed of driving a Formula 1 car? Like a Cessna pilot fantasizing about flying an F-15 or a baseball fan imagining knocking one of Orel Hershiser's pitches out of the park, most driving enthusiasts look to F1 as the ultimate adventure. And it's more than just the thought of driving one of the world's highest-performance machines. The high-brow ambiance, the jet-set lifestyle, and the sheer excitement that surround Formula 1 are at least as tantalizing as the driving.

Alas, the dream is shattered by even the briefest rational reflection. Realistically, none of us will ever race in a Grand Prix. In fact, few of us stand a chance of even *sitting* in a Formula 1 car. Reaching the top rung of motor racing requires a combination of talent, money, and commitment that only a handful of individuals possess.

There is, however, a way—albeit a costly one—of sampling the Formula 1 experience. For $97,000, Ferrari will put you into a Mondial t Cabriolet, complete with open-air bodywork, the prancing-horse emblem worn by more than 100 Grand Prix–winning cars, a shrieking engine, a Formula 1–inspired gearbox, a mid-engine chassis with sophisticated suspension and brakes, and enough charisma to make you the center of attention wherever you go.

Not only does the Mondial t Cabriolet offer all the right pieces, it tingles your soul with all the right sensations. Slide into the cockpit and you're surrounded by purposeful gauges and controls. Turn the key and you light off a hot-blooded powerplant that can't wait to be unleashed. Stand on the throttle and you are shoved back by smooth, hard thrust that never seems to quit. Driving the Mondial t is definitely not your average homogenized automotive experience. What we have here is your basic Formula 1 street machine.

The engine is so strong that our Mondial t Cabriolet test car, actually a 1989 model, pulled to its 7500-rpm red-

PHOTOGRAPHY BY DAVID DEWHURST

The Mondial is powered by Ferrari's largest V-8, the same longitudinally mounted, 300-hp, 3.4-liter unit fitted to the new 348.

line in fifth gear—an honest 159 mph. That's faster than any previous Mondial we've tested.

Acceleration is also impressive. Although the Mondial t is a bit sluggish off the line—even with a good, tire-smoking start—it reaches 60 mph from rest in 6.2 seconds and runs the quarter-mile in 14.5 seconds at 98 mph. Keep the throttle down and the Ferrari really comes into its own: it needs but 32.5 seconds to reach 130 mph. The Mondial t can't quite beat a Corvette or a 300ZX Turbo across an intersection, but it will pull away steadily from both cars at anything above an urban crawl.

Motivating the 3540-pound Mondial t is Ferrari's newest engine: the same longitudinally mounted, 32-valve, 3.4-liter V-8 fitted to the spanking-new 348. Developing 300 hp at 7000 rpm and 229 pound-feet of torque at 4000, this dynamic energy cell makes electrifying sounds that you don't expect to hear from a car wearing license plates.

Like its Ferrari V-8 predecessors, the 3.4-liter engine uses a flat crankshaft that produces perfectly even firing intervals between its two cylinder banks. This design makes for efficient exhaust tuning—at the expense of greater vibration than in conventional V-8 configurations. It also contributes to a high-rpm shriek that makes the Corvette's V-8 burble and the 300ZX's turbocharged hum sound positively wimpish. Wind the Mondial t to the redline and you hear 90 decibels' worth of precisely meshing gears, harmonically tuned intake pulsations, and staccato exhaust explosions. Blast through a tunnel with the top down and you can easily imagine that you're accelerating toward the harbor in the Monaco Grand Prix.

Unfortunately, when it's time to change gears the Mondial t brings you back to reality. No Grand Prix car could be so hard to shift. Despite its all-new design, the transverse gearbox feels as unwieldy as its predecessors. It's stiff, sticky, and generally uncooperative. The gated shift lever responds best to a heavy

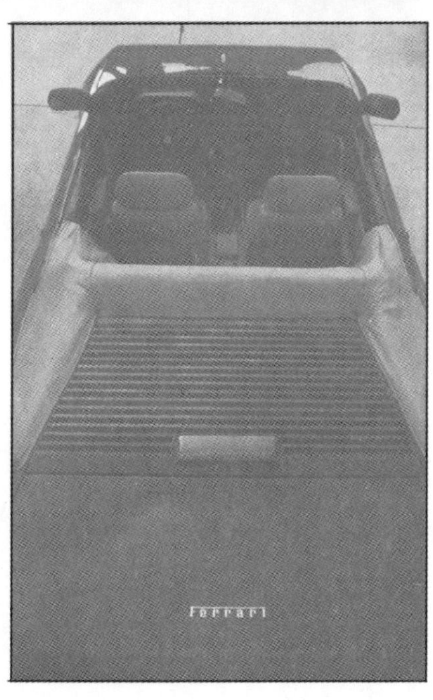

hand—although when the transmission is cold, you fear that the chrome shift lever will bend before the transmission moves into second gear. This is not the sort of gearbox you want to be stuck with in heavy traffic.

In other respects, the Mondial t is perfectly at home in an urban environment. The power-assisted steering makes low-speed maneuvering easy, and the suspension glides over most of the pockmarks and undulations that mar much of our transportation infrastructure.

Some of the credit for the excellent ride belongs to the Mondial t's electronically adjustable shock absorbers. The system offers three settings; within each, the damping varies with changes in speed and longitudinal and lateral acceleration. Even in its plushest mode, the system tightens up the shocks sufficiently to provide excellent control when you pick up the pace. In fact, we found little reason ever to use the stiffer settings—except to heighten the car's Formula 1 character.

The direct steering also contributes to the sporting spirit. The power-assisted rack-and-pinion mechanism provides a light and delicate touch with excellent feel. And that's important, because the revamped Mondial t maintains the handling characteristics of the previous mid-engined V-8 models. As you start cornering hard, you first encounter reassuring understeer. But as you approach the car's 0.85-g cornering limit, the tail steps out—and then you're in for a challenging ride. Unlike a Porsche 911, whose lift-throttle oversteer can be controlled by firmly reapplying the throttle, the Mondial t steps out with no predictable way to pull the tail back into line. Basically, you fight the wheel like crazy until you somehow gather the car up.

This tail-happiness makes fast driving challenging. Indeed, the Mondial's pendulumlike limit handling keeps it from being as quick over the road as many lesser sports cars. But it also provides an opportunity to sample firsthand the excitement that is the Grand Prix driver's stock in trade.

We find little to fault with the Mondial t's brakes. The anti-lock–equipped four-wheel discs stop the car from 70 mph in just 176 feet, and they resist fade even after repeated braking from high speeds.

Good as the Mondial t Cabriolet is in providing a simulated Formula 1 experience, it is also a remarkably practical package. In coupe or cabriolet form, the Mondial t is the most comfortable Ferrari on the market today. The back seat is too small for adults, but it does provide a handy place to toss your briefcase and overcoat. And the additional interior volume provides a spaciousness lacking in more closely coupled two-passenger sports cars.

A new dashboard and interior layout make the Mondial t airier than previous models, but the arrangement of the instruments and controls is still haphazard. Although the steering column's tilt is adjustable, it's impossible to find an angle that offers both a good hand position *and* a decent view of the instrument cluster. Happily, the wheel is within easy reach, and the overall driving position is reasonably comfortable.

Raising or lowering the Cabriolet's top is relatively easy. You must perform a number of simple maneuvers in the proper sequence to have any hope of getting the top into its well; but once you master the drill, the only tedious task is the fitting and removal of the boot. With the top up or down, our blood-red Mondial t looked smashing.

Admittedly, the Mondial t Cabriolet provides only a taste of the Grand Prix experience. But it's a stronger flavor than almost any other car can provide. Plus, the Mondial t is easily the most comfortable and practical of the high-priced Italian exotics.

You can even drive it without wearing flameproof underwear. ●

NSX vs 348ts

SHOOTING FOR THE TOP

BY KIM REYNOLDS

PHOTOS BY JOHN LAMM & DAVID W. BIRD II

DEC
CALIFORNIA
3421
STRIBUTOR

A few years ago, before Australian tycoon Alan Bond lost most of his money, he purchased Vincent Van Gogh's *Irises* and exhibited it all around Australia. Subsequently, it was rumored that what the public actually saw was a reproduction of the masterwork, displayed instead of the real thing for security reasons. Whether this is true or not, what makes the original worth $30 million or so, and a terrific phoney almost contemptible, is originality.

Of course, Ferraris are more plentiful than Van Goghs, but the mystique of the Prancing Horse's sports, GT and racing cars is carved just as deeply into the imaginations of automotive enthusiasts. With two seasons of Grand Prix domination behind it, Honda is feeling confident that in its Acura NSX, it can challenge Ferrari's newest, the 348, regardless of the fact that the road to Maranello is littered with everything from Pegasos to Panteras that aspired to Ferrari's status and fell by the wayside.

In typical Japanese fashion, it's a challenge that's also impressive for its car per dollar value. You can hear Honda's management challenging, "If we can engineer, build, and sell the NSX for $55,000, why does your 348 cost $97,000?" After all, Ferraris are no longer built by bib-overalled Giuseppes and Luigis with cigarettes in the corners of their mouths, tapping out aluminum bodywork on wooden bucks. The production lines on which the 348s are as-

■ VTEC stands for Variable Valve Timing and Lift Electronic Control System. What it means is that the NSX uses two different intake cam profiles to improve engine flexibility and still produce 270 bhp from the normally aspirated V-6 engine.

sembled are as bright with neon lights and loaded with automation as just about anyone's.

Dimensionally, the Acura NSX is 6.7 in. longer and 3.3 in. narrower, stands equally high (or low) and rides on a 3.1-in.-longer wheelbase than the Ferrari. Honda's stylists have said that when the NSX's shape was drawn, they had General Dynamics' F-16 in mind. And that's evident in its forward-positioned, bubble-like greenhouse, long tail and kicked-up fin supports for the rear wing. They may have also been thinking about the Testarossa's side air intakes and the Porsche 959's tall, free-standing rear wing. The NSX's low prow, a face full of features and beveled creases all its own, still somehow lacks a sharp identity, always looking like you've seen it somewhere before. In fairness, the NSX is an eye-catching, voluptuous whole, but it's detailed with alternately generic and derivative elements that leave you hungry for Giugiaro or Gandini . . . or Pininfarina.

Although the 348's body (by Pininfarina) is itself derivative, this time all the elements come from the same well. Following on the popular 328, the 348 appears more compact and muscular, impressions justified by a comparison of their dimensions, which identify the newer Ferrari as shorter and considerably wider. It's also less stricken with Coke-bottle-waist—probably doctored out of the design as much by aerodynamic concerns as changing fashion. Like the rest of the shape, the 348's nose is tauter and more squared-off than the NSX's, though recognizably a Ferrari by its grinning egg-crate grille. However, don't look too closely at it, because it's entirely nonfunctional—a surprising note of fakery in a body shape that's otherwise impressively responsible to both form and function. The engine cooling air, of course, is respirated through the ubiquitous slatted side ducts, as introduced on the Testarossa.

■

The 348's seats are set inboard 13 in. (2 in. more than the NSX's) but, once fully ensconced behind the 348's wheel, the environment has a lean, athletic tone. Our test car was a ts version (that is, a spyder, though targa would be more accurate), in which the pop-out top stores vertically behind the seats. This consumes about 2 in. of seat travel, noticed by several taller drivers. And the wide of beam grumbled as well, complaining of being perched atop the seats instead of fitting into them, though slim builds will find excellent support. Overall, the 348's interior deserves praise for being a big step in the right direction for Ferrari ergonomics, particularly the better organized switch arrangements, large air vents and better outward vision. Still, glitch-

■ The Ferrari's engine contains 300 horses. Raucous and loud where the NSX engine is strangely quiet, both provide exotic thrills.

■ **For all its swoopy, futuristic styling, the NSX sports a surprisingly contemporary interior, not that far removed from that in the Legend. The tool kit is firmly held in place, for those exuberant handling maneuvers this car encourages.**

es abound, such as the adjustable steering column that pivots near your knees, causing the wheel's rim to alternately interfere with bigger driver's thighs or obscure the upper portion of the instrument pod. And worse are problems with the pedals, from their severe offset to the right (the clutch is almost in line with the steering column), to the throttle pedal's clumsy floor-mounted pivot box that constantly feels like something has rolled under your heel.

Perhaps the greatest challenge to driving the NSX is ducking the low-cut door opening while entering. But from there on, it's truly the ergonomic exotic. Forward vision is terrific from the cab-forward driving position, and even glances over the shoulder in traffic are rewarded. The pedals are only slightly shifted to the right due to wheel-well intrusion, and an easy flick of the ankle is all that's needed for quick heel and toe blipping. The shifter works in butter-smooth short, snick-snick throws and, being a Honda product, it goes without saying that all of the controls, switches and knobs are logically placed and simple to use. Stitched leather covers the flowing shapes of the dash, door panels and widely accommodating seats in a relaxed, almost draped manner, particularly compared to the Ferrari's look of leather that is seemingly vacuum-formed to its interior's myriad oblique and muscular surfaces.

■

Of the two car's mechanical architectures, the NSX's follows common mid-engine practice, having a sidewinder engine and transaxle, as did the Lamborghini Miura and Ferrari's 8-cylinder cars until the 348. Up front are the radiator and space-saver spare tire; at the stern is a well-insulated and reasonably commodious trunk.

The 348, however, adopts the Testarossa's scheme of condensing the powertrain and its ancillaries into one tight package, meaning the radiators have been repositioned just ahead of the rear wheels. Consequently, the rear of the car is consumed by machinery, leaving the front available as a surprisingly large but awkwardly accessible trunk. And where's the spare? Ferrari says 348 drivers who suffer a flat tire aren't interested in changing it themselves, so none is fitted in the car.

The Ferrari's 300-bhp, 90-degree, 3.4-liter V-8 engine is essentially another enlargement of the 3.2-liter that powered the 328. Composed of an aluminum block and head with Nicasil-coated wet-steel liners, double overhead cams and four valves per cylinder, it breaks no new ground in Ferrari engine philosophy.

However, unlike the chasis on the 308 and the 328, the 348's chassis carries the unit longi-

tudinally, with its power traveling along a short shaft over the gearsets to a remote flywheel and clutch at the stern. There, bevel gears then turn the power 90 degrees and feed it into the transverse transmission.

With a 10.4:1 compression ratio, the Ferrari engine's output of 88.1 bhp/liter is impressive, but Honda's 90.7 bhp/liter from a 10.2:1 compression ratio is wizardly. Honda's 90-degree all-aluminum 3.0-liter V-6 produces 270 bhp at 7100 rpm and easily screams up to its 8000 rpm indicated redline. Titanium connecting rods, a production car first, are significant in allowing such high engine speeds for the NSX. Helping aspirate the cylinders over such a large rev band are two important intake tract technologies. First, is the variable-volume intake manifold which, beginning at 4800 rpm, couples the main plenum to an auxiliary chamber for better induction resonance. Then, between 5800 and 6000 rpm, a second system, called VTEC, further improves high rpm breathing. Meaning Variable Valve Timing and Lift Electronic Control System, it allows each cylinder's four valves to follow individual lobes below 5800 rpm; beyond that, it hydraulically locks their cam followers together tracking more radically profiled lobes.

Ferrari has little reputation for chassis and suspension inventiveness. However, the 348 breaks some new ground for Maranello in being Ferrari's first unit body, though an attached subframe carries the drivetrain and rear suspension's double A-arms (also used at the front). Weighing in at 3270 lb, the car's weight was reduced by an aluminum hood and rear deck,

but its 40/60 front/rear distribution is noticeably more biased than the previous 328's 44/56.

This was driven home by some twitchiness through our slalom cones and an easy willingness to do a half loop if the throttle was released circling the skidpad. However, driven with due respect on public roads, the 348 has a hard-edged, no-nonsense demeanor that's emphasized everywhere from its heavy steering to its firm ride and stiff, mechanical shift action. Press the 348 a little harder into a corner, modulating your throttle opening and steering angle, and images of Nigel Mansell on a fast lap fill your imagination . . . and create a little smile that the 348's character is strong enough to turn grown men into make-believe racing heroes.

In comparison to the Ferrari, the NSX's controls feel gossamer light: Steering, shift and clutch all need no more than an easy touch. Much of this has to do with the chassis' remarkably light weight, all-aluminum unit construction, which is another mass production first. And though the NSX's weight distribution is only slightly better balanced than the 348's, at 41/59, none of the Ferrari's opposite-lock bravado is ever required, even tossed through the most severe transitions.

At the rear, the NSX is suspended by conventional double A-arms, but up front, similar A-arms have their forward bearings joined to an additional component Honda calls a compliance pivot. In action, sharp bumps that can cause fluctuations in toe angle are constrained to rotate the compliance pivots to a predetermined degree, compensating for the wheel's anticipated toe-deflection.

If even more handling security is desired, Honda's traction control system, TCS, is a button's push away. The system analyzes any differential in front wheel speeds (provided by their ABS sensors), together with steering wheel angle and road speed, and quickly determines if a loss of traction (or control) is imminent. If trouble is sensed, the engine's power is appropriately reduced.

The NSX is a car of dual natures. Its ride is both tautly controlled on smooth roads and surprisingly compliant on poor ones. The NSX's handling is razor-precise driven quickly yet not at all nervous at lesser speeds. And most strikingly, at idle, the engine gently whispers, though becoming a shrill, almost jet-like whine under full throttle.

Has Honda bettered Ferrari? Even ignoring their price differences, virtually every driver who spent time behind their wheels felt the NSX would be the better choice as a sports car to live with. Its advantages in comfort, outward vision, ergonomics and lighter control effort are that significant. However, most of these drivers also agreed that the Ferrari would be their pick for the sheer fun of a Sunday morning's dash down a favorite mountain or backroad. Let's say the 348 is the better exotic, while the NSX is the more successful car.

That conclusion is somewhat predictable. Breaking into the exotic car market in 1990, as Honda has done, means the world's critics are not going to allow many missteps. Yet what makes some of the world's great cars memorable are their imperfections, often originating decades earlier and to which we've gradually grown accustomed. Then there's the often stated observation that Ferrari really just builds the same car over and over, but each time it's a little better. Certainly we have in the NSX an automobile of highly refined exotic engineering, while in the 348, one of equally refined romance. Fortunately, there's room for both. ■

■ The 348's side strakes are obviously a derivative of the Testarossa. Not that surprising, when you consider that both are products of Pininfarina's styling house.

ACCOMMODATIONS

	Acura NSX	Ferrari 348 ts
Seating capacity	2	2
Headroom	36.0 in.	35.5 in.
Seat width	2 x 19.5 in.	2 x 18.5 in.
Front leg room	45.0 in.	43.5 in.
Trunk space	6.5 cu ft	6.8 cu ft

FUEL ECONOMY

	Acura NSX	Ferrari 348 ts
Normal driving	est 23.0 mpg	est 16.0 mpg
EPA city/highway	est 21/31 mpg	13/19 mpg
Fuel capacity	18.5 gal.	23.2 gal.

INTERIOR NOISE

	Acura NSX	Ferrari 348 ts
Idle in neutral	48 dBA	66 dBA
Constant 70 mph	71 dBA	76 dBA

ACCELERATION

	Acura NSX	Ferrari 348 ts
Time to speed		
0–30 mph	1.9	2.2
0–60 mph	5.7	6.0
0–80 mph	9.6	9.8
Time to distance		
0–100 ft	2.9	3.0
0–500 ft	7.7	7.9
0–1320 ft (1/4 mi)	14.0 sec @ 100.0 mph	14.3 sec @ 98.5 mph

PRICE

	Acura NSX	Ferrari 348 ts
List price, all POE	$60,000	$94,800
Price as tested	$60,000	$94,800

Acura NSX: Price as tested includes std. equip. (auto climate control, AM/FM stereo/cassette, ABS, traction control, leather int, central locking, cruise control; elect.-operated window lifts, mirrors & seats).

Ferrari 348 ts: Price as tested includes std. equip. (air cond, leather int, ABS, fitted luggage, central locking; elect.-operated window lifts & mirrors).

GENERAL DATA

	Acura NSX	Ferrari 348 ts
Curb weight	2985 lb	3270 lb
Test weight	3135 lb	3370 lb
Weight dist, f/r %	41/59	40/60
Wheelbase	99.6 in.	96.5 in.
Track, f/r	59.4 in./60.2 in.	59.2 in./62.2 in.
Length	173.4 in.	166.7 in.
Width	71.3 in.	74.6 in.
Height	46.1 in.	46.1 in.

ENGINE

	Acura NSX	Ferrari 348 ts
Type	dohc 4-valve V-6	dohc 4-valve V-8
Displacement	2977 cc	3405 cc
Bore x stroke	90.0 x 78.0 mm	85.0 x 75.0 mm
Compression ratio	10.2:1	10.4:1
Horsepower, SAE	270 bhp @ 7100 rpm	300 bhp @ 7000 rpm
Torque	210 lb-ft @ 5300 rpm	229 lb-ft @ 4000 rpm
Maximum engine speed	8000 rpm	7500 rpm
Fuel delivery	Honda PGM-FI elect. port	Bosch Motronic M2.5 elect. port
Fuel	premium unleaded, 91 pump oct	premium leaded, 91 pump oct

BRAKING

	Acura NSX	Ferrari 348 ts
Minimum stopping distance		
From 60 mph	128 ft	144 ft
From 80 mph	220 ft	250 ft
Control	excellent	excellent
Pedal effort for 0.5g stop	17 lb	20 lb
Fade, effort after six 0.5g stops from 60 mph	17 lb	20 lb
Brake feel	excellent	excellent
Overall brake rating	excellent	excellent

HANDLING

	Acura NSX	Ferrari 348 ts
Lateral accel (200-ft skidpad)	0.90g	0.91g
Balance	mild understeer	neutral
Speed thru 700-ft slalom	64.8 mph	62.8 mph
Balance	neutral	neutral

Subjective ratings consist of excellent, very good, good, average, poor; na means information not available.

CHASSIS & BODY

	Acura NSX	Ferrari 348 ts
Layout	mid engine/rear drive	mid-engine/rear drive
Body/frame	unit aluminum	steel & aluminum/unit steel & skeletal steel
Brakes, f/r	11.1-in. vented discs/11.1-in. vented discs, vacuum assist ABS	11.8-in. vented discs/12.0-in. vented discs, vacuum assist, ABS
Wheels	forged alloy, 15 x 6½J f, 16 x 8J r	alloy, 17 x 7½ f, 17 x 9 r
Tires	Yokohama A-022, 205/50ZR-15 f, 225/50ZR-16 r	Bridgestone RE71, 215/50ZR-17 f, 255/45ZR-17 r
Steering	rack & pinion	rack & pinion
Turns, lock-to-lock	3.2	3.0
Turning circle	38.2 ft	39.2 ft
Suspension, f/r	upper & lower A-arms, compliance pivot, coil springs, tube shocks, anti-roll bar/upper & lower A-arms, coil springs, tube shocks, anti-roll bar	upper & lower A-arms, coil springs, tube shocks, anti-roll bar/ upper & lower A-arms, coil springs, tube shocks, anti-roll bar

DRIVETRAIN

	Acura NSX		Ferrari 348 ts	
Transmission	5-sp manual		5-sp manual	
Gear	Ratio	(Rpm) Mph	Ratio	(Rpm) Mph
1st	3.07:1	45	3.21:1	47
2nd	1.73:1	80	2.11:1	72
3rd	1.23:1	112	1.46:1	104
4th	0.97:1	142	1.09:1	138
5th	0.77:1	est (7510) 168	0.86:1	est (7290) 171
Final drive ratio	4.06:1		3.56:1	
Engine rpm @ 60 mph in 5th	2680		2560	

MISTER MEETS MASTER

GREAT MASTERS HAVE ALWAYS BEEN COPIED AND THE CAR INDUSTRY IS NO EXCEPTION. GREG KABLE COMPARES THE FERRARI-INSPIRED MR-2 WITH THE REAL McCOY

Okay, okay. I know what you're thinking . . . and you're right. Toyota's all-new, better-looking and more potent MR-2 is clearly not a serious challenger to Maranello's latest and finest, the striking Ferrari 348tb. We never said it was.

In fact, at no time during our early office debate, aimed at solving the perplexing question of finding a suitable competitor for Toyota's so-called small Ferrari, was the 348tb even mentioned. Initially we were too preoccupied in fielding questions on the relevance of cars such as the four-wheel steer Honda Prelude and Mazda MX-6, as well as the multi-award winning Mazda MX-5. On reflection, we realised that, although these cars might display some sort of cost competitiveness, in an overall sense they just aren't in the same league.

Then, just one week before we were due to take delivery of the pre-production MR-2 for evaluation (before it made its debut in local

showrooms) it dawned upon us. If Toyota has built its own small Ferrari, as its senior officials have no hesitation in claiming, what better way of confirming this than getting the two together for a back-to-back comparison. Italy versus Japan, if you like.

The aim was to discover just how successful Toyota, the world's third-largest car maker, is at producing a low-cost alternative to what's been described as the world's best production sports car — the Ferrari 348tb (tb denoting transversal berlinetta).

When it lands here in September, the Ferrari 348tb will bring with it a price-tag nearly six times that of the new MR-2. But is price the only difference?

Here in the office we liked to argue it wasn't, but when I collected the glistening metallic-blue MR-2 from Toyota's Taren Point headquarters in Sydney I began to wonder if it was. Initial impressions of Toyota's new sporty were high. It felt taut, dynamic, looked good and turned heads. It possessed the hallmarks of a well-sorted sports car, with all the blemishes of the old model ironed out.

Then the 348tb made an appearance and, quick as I was to respond to the outstanding qualities of the MR-2, I was stunned by the mere presence of the Ferrari.

As a motor noter you get to drive some fairly impressive machinery from time to time, but rarely does the opportunity arise to sample one with such outstanding credentials as the Ferrari 348tb.

The Ferrari we were supplied with has a long history. It's the car that sat resplendent at the Frankfurt motor show in October last year for its world debut. From there it was flown to the US to feature in the Washington motor show and do the rounds of the US motor magazines. It was subsequently transported back to its motherland, Italy, for a number of on-going repairs before making its way Downunder to take part in the Adelaide Grand Prix festivities early in November.

It isn't until you view the new MR-2 in the flesh that you gain a full appreciation of its new style. With opinionated tastes some describe it as "breathtaking", while others label it a "bomb". For mine, Toyota's stylists have produced a package that's nicely proportioned, but which, when placed side by side with the 348tb, appears soft and doesn't possess the substance instilled in the unmistakable Pininfarina lines that adorn the Ferrari. The dramatic strakes that run the length of the doors, as well as those at the rear-end of the low and extremely wide 348tb, highlight its purpose, reminding you of the Maranello company's flamboyant tradition of creating not only the most potent but most imposing sports cars in the world. Aerodynamically the be-spoilered MR-2 comes out on top with a drag co-efficient of just 0.31, but despite its blunter front-end, squarer lines and much wider stance the 348tb is not far behind at 0.32.

For all its sweet sonics, the Ferrari isn't the least bit temperamental. It's as happy at 1000 rpm as it is at 7000 rpm. Run to high revs and it just beckons for more

Unlike its predecessor, the new MR-2 is definitely no pocket-rocket. At 4180 mm and 2400 mm respectively, it has grown appreciably and is just 50 mm shorter than the 348tb in both overall length and wheelbase measurements. The primary aesthetic difference is in width and height. At 1894 mm and 1170 mm, the squat 348tb is a considerable 194 mm wider and 70 mm shorter than the MR-2.

While on the outside both the 348tb and MR-2 possess similarities, under their elegant skins they're as different as Bach and the Beatles. The Ferrari's largely reworked quad-cam, 32-valve, all-alloy, 90-degree, V8 powerplant, with a total displacement of 3405 cc, is now mounted longitudinally rather than transversely, as in the MR-2. This allows enough space to accommodate repositioned side-mounted radiators as well as a new air-intake and exhaust system and a neatly engineered, unique, transversely mounted gearbox, similar in concept to that used in the Ferrari 312T F1 cars back in the late '70s, which is mounted aft of the block.

This combination of state-of-the-art technology produces a remarkable 221 kW at 7200 rpm and 323 Nm of torque at 4200

rpm. Substantially more than the MR-2, but what you have to remember is that the 348tb holds a 1.4-litre advantage.

The Ferrari is intended to be a sports car of unparalleled qualities. The sound of its V8 confirms this. It's quiet down low and from about 4000 rpm there's a definite lift in tone with all 32-valves chattering and the exhausts bellowing. But for all its sweet sonics, the Ferrari isn't the least bit temperamental. It's as happy at 1000 rpm as it is at 7000 rpm. Run to high revs in any gear and it just beckons for more, reflecting a character that's rarely found in any car.

Due to a slight gremlin in the 348tb's gearbox, caused when it was driven with insufficient transmission lubricant during an earlier test, we were unable to strap our electronic timing equipment to it. All the same, the factory claims it sprints from 0-100 km/h in just 5.6 seconds and covers a standing kilometre in 24.8 seconds. That's almost as fast as the Testarossa and, judging by the way it flattens you against the seat backs, is believable. Unfortunately, at the time of writing a standing 400 metre time was unavailable, but if reliable sources are correct it should come close to equalling the Testarossa's astonishing time of just 14.20 seconds.

Where the 348tb dazzles, the MR-2 impresses. The MR-2's compact transversely mounted 3S-GE engine is a hard worker. From as few as 1000 rpm in third gear the ever-willing twin-cam, 16-valve, 2.0-litre unit unfailingly responds without hesitation. Tractability is impressive right the way through to its 7300 red line. While it doesn't possess the bullet-like performance of the Ferrari, it is still an extremely fast and very exhilarating car to drive. A new twin-pipe exhaust system provides a gutsy and rasping exhaust note above 4000 rpm that, far from being a bugbear, serves only to remind you of its sporty nature. The gear-shift and long-reaching clutch, unlike others in the Toyota range, are well weighted as they should be in any true sports car.

The move to the punchier 3S-GE powerplant has provided the MR-2 with tremendous straight-line performance. With 117 kW at 6600 rpm and 190 Nm of torque at 4800 at its disposal, the weighty 1215 kg Toyota dashes from 0 to 100 km/h in 7.6 seconds. Two less than its predecessor and two more than the 348tb. A standing 400 metres is traversed in 15.35 seconds, outstanding for a naturally aspirated car of just 2.0-litre capacity.

Apart from the MR-2's general configuration and, of course, name, the only other feature that bears any remote resemblance at all to the old model is its fully independent suspension system.

Initial impressions of Toyota's new sporty were high. It felt dynamic and looked good. It possessed the hallmarks of a well-sorted sports car

Though even here the new car's larger dimensions have brought about a new, much more rigid, MacPherson strut set-up, with increased diameter arms and linkages as well as new mounting points and uprated coil springs.

With increased power the MR-2 also gets bigger stoppers all round. Ventilated discs now serve at each corner, backed-up by four piston calipers in the sharp end. Stopping is one the things the MR-2 does best. Pedal effort is exemplary and easy to modulate under all conditions.

Obviously a move by Toyota to limit high-speed twitchiness, the new MR-2 now gets differing rubber at each end. Up front there are 195/60 R14s on 6.0J X 14 rims while at the rear are 205/60 R14s on 7.0J X14 rims.

The 348tb, as you'd expect for a car with such race-bred qualities, boasts an impressive under-carriage. The suspension, as in the older 328 GTBi model, still employs unequal length control arms both at the front and rear, but the front now gets a more compact set-up. Rubber is an impressive 215/50ZR 17 on a 7.5J x 17 inch rim at the front and substantial 255/45ZR 17 on 9.0J x 17 at the rear. The Ferrari's five-spoke alloy wheels, like those on the Toyota, look impressive and purposeful.

Race tracks don't tell you everything you need to know about a car that's built for the road. But as part of the evaluation process, they do allow you to work up speeds and experiment with

Top left: **The leather-clad 348tb's interior is extremely functional and has an open-air feel.**
Bottom left: **An intrusive transmission tunnel mars the otherwise good interior of the MR-2.**
Top right: **The majestic Maranello 32-valve, 3.4-litre V8 powerplant propels the 348tb from 0-100 km/h in just 5.6 seconds.**
Bottom right: **Toyota's super-responsive 16-valve, 2.0-litre engine provides a gutsy and rasping exhaust note above 4000 rpm.**

on-the-limit dynamic behaviour in ways that would be dangerous, not to mention irresponsible and outright stupid, on public roads.

Finding the 348tb's ultimate potential would be hard. Its grip is just phenomenal. Over the open stretches of Sydney's Oran Park it simply glided around corners at speeds that would have had any lesser car spinning off into the distance. It's not easy to drive; the gearbox and low-speed steering demand more muscular effort than any other car I know. But in hindsight, that in itself is all part and parcel of what is commonly referred to as the Maranello mystique. It just wouldn't be a true Ferrari if it were easy to master.

As a handling benchmark, the 348tb has few contemporaries and the MR-2 isn't one of them. That's not to say the Toyota handles poorly. To the contrary. It flows beautifully through corners, the extra horsepower, new suspension and wider rubber combining to lift cornering speeds appreciably. It's much more predictable on the limit than the older model and in most cases you can punch all its 117 kW without worrying too much about where the rear-end will be upon exit. Like its predecessor, the new MR-2 is a true devil, tempting you to stomp the accelerator in situations you'd otherwise never consider. It's hard to bring to mind any other affordable sports car that combines such compromising ride and high-speed cornering ability. Both spring rate and damping verge on passenger car levels, yet respond sweetly when called upon in demanding and on-the-limit situations.

The Ferrari's black leather-trimmed seats, although devoid of substantial padding like the MR-2's, are supportive and provide substantial fore-and-aft adjustment which, together with the adjustable steering wheel, negates any need for the driver to assume the extreme, arms-fully-stretched, Italian-style driving position. Needless to say, the pedals are ideally arranged for heel-and-toe operation.

The newly designed wrap-around interior is extremely functional; its familiar red-on-black Veglia Borletti instruments, including speedo, tacho, oil pressure and water temperature gauges, stare out at you from behind the neatly leather-trimmed three-spoke Momo steering wheel. Matching fuel and temperature gauges, just a sideways glance away, are mounted in the central console that also houses controls for fog lights, radio/cassette and air-conditioning. The stubborn-shifting but beautifully crafted, chromed gear-lever and external gate remain a central beauty spot.

Before switching from the Toyota to the Ferrari I considered the former to possess a well integrated interior and, gripes about its intrusive transmission tunnel aside, I still do. The cluttered driving environment of the older MR-2 is gone, replaced by an entirely new, more free-flowing design that's pleasant to inhabit but which lacks the open-air feel of the 348tb.

The MR-2 caters for the enthusiast with large, some may say oversized, white-on-black instruments, including speedo, tacho, fuel and temperature gauges. The

Ferrari 348tb
3.4-litre, 5-speed manual

ENGINE

Location	mid-rear, long mounted
Cylinders	90° V8
Bore x stroke	85.0 x 75.0 mm
Capacity	3405cm^3
Induction	electronic multi-point fuel injection
Compression ratio	10.4 to 1
Valve gear	chain and belt driven dohc per bank, four valves/cyl
Power	221 kW @ 7200 rpm
Torque	323 Nm @ 4200 rpm
Maximum rpm	7500
Specific power output	64.9 kW/litre

TRANSMISSION

Type	5 speed manual
Driving wheels	rear

GEARBOX

Gear ratio		kmh 1000 rpm	Max Speed	At (rpm)
First	3.210	10.6	80	7500
Second	2.110	16.1	121	7500
Third	1.460	23.3	175	7500
Fourth	1.090	31.2	234	7500
Fifth	0.860	39.6	273	6900
Final-Drive	3.56			

SUSPENSION

Front	independent by unequal length control arms, coil springs and anti-roll bar
Rear	independent by unequal length control arms, coil springs and anti-roll bar
Wheels	alloy, 7.5J x 17 front 9.0J X 17rear
Tyres	Pirelli P700 215/50ZR 17 front 255/45ZR 17 rear

BRAKES

Front	ventilated 300 mm discs
Rear	ventilated 306 mm discs

STEERING

Type	rack and pinion
Turns, lock to lock	3.2
Turning circle	12.0 metres

DIMENSIONS

Wheelbase	2450 mm
Front track	1483 mm
Rear track	1577 mm
Overall length	4230 mm
Overall width	1894 mm
Overall height	1170 mm
Ground clearance	125 mm
Kerb weight	1393 kg
Weight to power	6.3 kg/kW
Fuel tank	91.0 litres

EQUIPMENT

Adjustable steering	yes
Air-conditioning	yes
Alloy wheels	yes
Central locking	yes
Clock	yes
Cruise control	no
Driver's footrest	yes
Intermittent wipers	yes
Oil pressure gauge	yes
Power steering	no
Power windows	yes
Radio/cassette	yes
Rear window wiper	no
Remote outside mirror adjust	2 elec
Sun roof	no
Tachometer	yes
Trip computer	no
Volts/Amps gauge	no

ACCELERATION

0-100 km/h	5.60 seconds
Standing 400m	n/a
Terminal speed (400m)	n/a

The above are averages of runs in opposite directions

Standing 400m, best	n/a
Terminal speed (400m), best	n/a

Figures by Ferrari.

LIST PRICE	n/a
PRICE AS TESTED	n/a
Includes options:	n/a

Toyota MR-2
2.0-litre, 5-speed manual

ENGINE

Location	mid-rear, trans mounted
Cylinders	four in-line
Bore x stroke	86.0 x 86.0 mm
Capacity	1998cm^3
Induction	electronic multi-point fuel injection
Compression ratio	9.5 to 1
Valve gear	chain driven single dohc per bank, four valves/cyl
Power	117 kW @ 6600 rpm
Torque	190 Nm @ 4800 rpm
Maximum rpm	7300
Specific power output	58.6 kW/litre

TRANSMISSION

Type	5 speed manual
Driving wheels	rear

GEARBOX

Gear ratio		kmh 1000 rpm	Max Speed	At (rpm)
First	3.285	8.5	62	7300
Second	1.960	14.2	104	7300
Third	1.322	21.1	154	7300
Fourth	1.028	27.1	198	7300
Fifth	0.820	34.0	221	6500
Final-Drive	3.94			

SUSPENSION

Front	independent by MacPherson struts with coil springs and anti-roll bar
Rear	independent by MacPherson struts with coil springs and anti-roll bar
Wheels	alloy, 6.0J x 14 and 7.0J x 14
Tyres	Bridgestone RE85VR front RE88VR rear 195/60 R14 85V front 205/60 R14 88V rear

BRAKES

Front	ventilated 258 mm discs
Rear	ventilated 263 mm discs

STEERING

Type	rack and pinion
Turns, lock to lock	3.7
Turning circle	9.8 metres

DIMENSIONS

Wheelbase	2400 mm
Front track	1470 mm
Rear track	1450 mm
Overall length	4180 mm
Overall width	1700 mm
Overall height	1240 mm
Ground clearance	n/a
Kerb weight	1215 kg
Weight to power	10.4 kg/kW
Fuel tank	55.0 litres

EQUIPMENT

Adjustable steering	yes
Air-conditioning	no
Alloy wheels	yes
Central locking	yes
Clock	yes
Cruise control	no
Driver's footrest	yes
Intermittent wipers	yes
Oil pressure gauge	no
Power steering	no
Power windows	yes
Radio/cassette	yes
Remote outside mirror adjust	2 elec
Sun roof	yes
Tachometer	yes
Trip computer	no
Volts/Amps gauge	no

ACCELERATION

0-100 km/h	7.60 seconds
Standing 400m	15.35 seconds
Terminal speed (400m)	142.1 km/h

The above are averages of runs in opposite directions

Standing 400m, best	15.10 seconds
Term speed (400m), best	146.2 km/h

Figures by Datron Correvit L3 digital electronic equipment.

LIST PRICE	n/a
PRICE AS TESTED	n/a
Includes options: air conditioning	n/a

fabric-trimmed seats are comfortable, extremely well bolstered and offer both height and lumber adjustment. The pedals are well laid out and equal the Ferrari's in every respect. Creature comforts include a top-rate sound system, although, surprisingly, our prototype test car did not boast air-conditioning.

During our drive program it was hard not to let your heart rule your head, but although the Ferrari evokes passions that no other car could, nothing can be taken away from the Toyota.

I'm in no way prepared to pronounce the Toyota MR-2 the best sports car in the world, but it is arguably the best sports car you can buy for its projected price tag of less than $40,000, no questions asked. Other sporties go faster and corner better, but few, if any at all, come close to matching Mister Two's sheer thrill for dollar value. It will do wonders for Toyota's image in every market in which it is offered.

Between these two vastly different cars there's no comparison, just some extremely interesting truths.

HERE'S A CAR that generates controversy in just about any circle. Even among Ferrari cognoscenti, some see the Mondial t Cabriolet as an effete concession to the times, hardly worthy of the Cavallino emblem. Other Ferrari folks, however, consider it the most useful car out of Maranello, and thus the most coveted in any realistic day-in-day-out sense.

Among enthusiasts in general, there are those who find being around a Ferrari, any Ferrari, a rare treat. Therefore, for them, the Mondial is an object of veneration. On the other hand, some of these enthusiasts are fortunate enough to sample a Ferrari, and a few come away disillusioned. What's left is an intellectual appreciation for the marque, but the mystique is a bit tarnished.

Last, among those of us here at R&T, there's everything from cognoscenti to newcomers, elitists to populists, wind-in-the-face types to those actually not liking Morgan trikes; can you imagine? So it's no wonder driving a Mondial t Cabriolet for a couple of weeks left its mark—lots of them, in fact.

Before sharing these, though, let's fit the car into the full Maranello spectrum. Its 1980 Geneva show introduction dates the Mondial Coupe as the oldest of current Ferraris; four years later came the soft-top Cabriolet of our road test interest here. Sharing its powertrain is the 348 introduced last year as replacement for the 308/328, each of these models sporting a targa version, thus leaving the Cabriolet the only truly open Ferrari these days. The exotic Testarossa and hyperexotic F40 complete the range.

By way of numbers, the 348 accounts for something like half of Ferrari North America's business; the Testarossa, around 30 percent; the F40, only a handful; and the Mondial rounds things out at roughly 20 percent, Cabriolets outselling Coupes by 9 to 1.

Do we infer from this that the Mondial is a rara avis or merely the lame duck of the Ferrari lineup?

"I really like the car's open configuration," one enthusiastic owner

FERRARI MONDIAL t CABRIOLET

Maranello's most practical product

PHOTOS BY GUY SPANGENBERG

Ferrari
ROAD & TRACK
R&T
ROAD TEST

told us, "and, besides, it's the only Ferrari I fit into." Indeed, the interior of the Mondial is the most spacious of current models.

Don't read too much into its 2+2 nomenclature, however. The second pair of seats is for occasional use at best, and then only if the front-seat occupants are generous in sharing their leg room. Used as a *bi-posti*, though, the car offers its driver and front-seat passenger ample luggage and storage space, well beyond that of any other Ferrari.

What's more, the Mondial's driving position doesn't require the exaggerated seat rake of more extreme examples of Pininfarina styling; even taller sorts can sit relatively erect. And the rest of the interior is characteristically Ferrari, which is to say a whole spectrum of good, bad and indifferent.

For instance, the oversize pedals are offset distinctly to the right; this, to clear the left front fender cutout. Some Ferrari newcomers find this awkward; others accommodate in a minute. The thickly wrapped steering wheel, despite its adjustability for rake, lamentably blocks interesting segments of the 7500-redline tach. By contrast, the 180-mph speedometer is in clear view, and this is the one that can keep you out of—or get you into—a lot of trouble.

The gear lever and its exposed, polished gate are also characteristically Maranello. And so are the opposing views as to its function. "Like all Ferrari boxes," noted a resident *tifosa*, "this one encourages the confidently firm hand motions that come with practice."

Countered a newcomer: "Call me a pragmatist, but large chunks of Ferrari mystique erode when I'm dealing with this cantankerous shift linkage."

Even he, though, fell prey to the wonderful noises emitted by the Mondial's 3405-cc 300-bhp V-8. "It's as though there's an extra set of gears driven off the camshafts whose sole purpose is to make beautiful sounds."

The t in our Cabriolet's name is for *trasversale*, indicating the transverse orientation of its 5-speed transmission behind the longitudinal amidships V-8, a la Ferrari Formula 1 practice. As opposed to earlier Mondials in which the engine was transverse, this powertrain layout offers a lower center of gravity, most evident in the powerplant's relatively low residence in the engine bay.

Mondial performance has traveled an upward curve since the first *quattrovalvole* appeared in 1984, to the point that the current t Cabriolet gets full marks. It accelerates from a standstill to 60 mph in just 6.6 seconds, reaches the quarter mile in 15.0 sec. at 93.0 mph and is capable of topping out at 154 mph. Its oversize disc brakes, now with ABS, haul the car down from 60 mph in 147 ft. On our skidpad, the Mondial turned in a lateral acceleration of 0.89g.

■ Veglia gauges, with their characteristic orange markings, grace the instrument panel. Quality of the interior's leatherwork is exceptional; the downside of an open car is that the aroma of leather gets away!

FERRARI MONDIAL t

0–60 mph	6.6 sec
0–¼ mi	15.0 sec
Top speed	est 154 mph
Skidpad	0.89g
Slalom	na
Brake rating	very good

PRICE

List price, all POE .... **$96,300** Price as tested **$96,300**

Price as tested includes std equip. (air cond, leather interior, elect. window lifts, elect. adj mirrors, central locking, ABS).

ENGINE

Type	dohc 4-valve/cyl **V-8**
Displacement	208 cu in./3405 cc
Bore x stroke	3.35 x 2.95 in./ 85.0 x 75.0 mm
Compression ratio	10.4:1
Horsepower (SAE):	**300 bhp @ 7000 rpm**
Torque	**229 lb-ft @ 4000 rpm**
Maximum engine speed	7400 rpm
Fuel injection	Bosch Motronic elect. port
Fuel	premium unleaded, 91 pump oct

GENERAL DATA

Curb weight	**3640 lb**
Test weight	3790 lb
Weight dist, f/r, %	44/56
Wheelbase	104.3 in.
Track, f/r	59.9 in./61.4 in.
Length	**178.5 in.**
Width	**71.3 in.**
Height	**48.6 in.**
Trunk space	5.0 cu ft

DRIVETRAIN

Transmission .... **5-sp manual**

Gear	Ratio	Overall ratio	(Rpm) Mph
1st	3.21:1	11.43:1	41
2nd	2.11:1	7.52:1	63
3rd	1.46:1	5.20:1	91
4th	1.09:1	3.88:1	121
5th	0.86:1	3.06:1	est (7400) 154

Final drive ratio 3.56:1

Engine rpm @ 60 mph in 5th 2880

CHASSIS & BODY

Layout	**mid engine/rear drive**
Body/frame:	steel, aluminum/skeletal steel
Brakes, f/r	**11.1-in. vented discs/ 11.0-in. vented discs;** vacuum assist, ABS
Wheels	cast alloy; **16 x 7J f, 16 x 8J r**
Tires	Goodyear Eagle ZR; **205/55ZR-16 f, 225/55ZR-16 r**
Steering	**rack & pinion**, power assist
Turning circle	38.9 ft
Suspension, f/r:	**upper & lower A-arms,** coil springs, tube shocks, anti-roll bar/**upper & lower A-arms,** coil springs, tube shocks, anti-roll bar

FUEL ECONOMY

Normal driving	est 15.0 mpg
EPA city/highway	12/19 mpg
Fuel capacity	22.5 gal.

INTERIOR NOISE

Idle in neutral	66 dBA
Constant 70 mph	77 dBA

ACCELERATION

Time to speed	Seconds
0–30 mph	2.3
0–60 mph	6.6
0–80 mph	11.0
Time to distance	
0–100 ft	3.1
0–500 ft	8.2
0–1320 ft (¼ mi)	15.0 @ 93.0 mph

BRAKING

Minimum stopping distance	
From 60 mph	147 ft
From 80 mph	258 ft
Control	excellent
Pedal effort for 0.5g stop	17 lb
Fade, effort after six 0.5g stops from 60 mph	17 lb
Brake feel	excellent
Overall brake rating	very good

HANDLING

Lateral accel (200-ft skidpad)	0.89g
Balance	neutral
Speed thru 700-ft slalom	na
Balance	na

Subjective ratings consist of excellent, very good, good, average, poor; na means information is not available

Refinements to the car include driver-adjustable shock absorbers, befitting the Ferrari heritage in ranging from reasonably firm to intermediately so to downright racer-stiff. Another refinement is power-assisted steering that actually improves what was already one of the world's better rack-and-pinion systems. Gone is the previous low-speed heaviness; what's more, an on-center dullness typical of earlier Ferraris is gone as well.

To many of us, it's less than completely satisfying to operate the Cabriolet's soft top. This one goes up quite easily; it's the lowering that demands patience and perseverance. The owner's manual even cites potential misalignments, something you would expect Ferrari's engineering expertise to have eliminated during development.

Its appearance is also paradoxical: The Cabriolet's top looks quite handsome when up, several staff members liking the way its tapered sails rakishly flank the engine hatch. But, when stowed, the top resides rather high, obstructing vision to the rear, making side mirrors invaluable and reminding one that this is not just any convertible—it's an exotic.

In fact, though, this duality of the practical and the exotic sums up the Mondial very well. Despite their quirks, Ferraris have earned enviable reputations as durable machines when maintained properly. And *because* of their quirks, these cars offer added satisfaction to those learning how to drive them properly. In particular, the Mondial's blend of Ferrari attributes, good and bad, sets it distinctly apart from other cars on the road, even other exotics. And the Cabriolet's 2+2 nature sets it distinctly apart, even from other Ferraris.

Test Notes . . .

- **Unlike earlier 2.9-liter models, the Mondial t's new 300-bhp 3.4-liter engine produces acceleration times worthy of a Ferrari. Driven fast or slow, the Mondial is exhilarating.**
- **The Mondial t's longitudinal engine/ transverse gearbox layout eliminates axle tramp under hard acceleration and considerably improves shifting.**

E L A B O R A T I O N

WITH MANY MAJOR MANUFACTURERS WANTING TO PRODUCE SPECIAL-BODIED, LIMITED-PRODUCTION CARS, THE COACHBUILDERS OF ITALY ARE MAKING A COMEBACK. GRAZIELLA DIANA FERRERO, OUR HOLY LADY OF ITALIAN MOTORING AND BLESSED ACCELERATION, LOOKS AT A ZAGATO FERRARI 348TB. THE STYLING ELABORATIONS WERE CAPTURED ON KODACHROME BY ROBERTO CARRER.

A CLICHE IN THE SPORTS CAR CLASSIFIED ads that's guaranteed to get you at least a few phone calls is to say your car "looks just like a Ferrari." Never mind which Ferrari it's supposed to look like, it's a great phrase. And as long as the car in question isn't betrayed by a production sedan profile, it can even be set on Volkswagen mechanicals. Hey, everyone wants to drive a car that looks just like a Ferrari. ¶ What we have here is full opposite lock: a Ferrari that's not supposed to look just like one. At least, not just like the 348tb it's based on. The owner of this car approached Zagato with his own definite ideas of what a Ferrari should look like, and the changes wrought to the body have been extensive enough that the car's exact heritage is not immediately evident.

ZAGATO
CO B10944

"WHAT WE HAVE HERE IS FULL OPPOSITE LOCK: A FERRARI THAT'S NOT SUPPOSED TO LOOK JUST LIKE ONE. AT LEAST, NOT JUST LIKE THE 348TB IT'S BASED ON. THE OWNER OF THIS CAR APPROACHED ZAGATO WITH HIS OWN DEFINITE IDEAS OF WHAT A FERRARI SHOULD LOOK LIKE, AND THE CHANGES WROUGHT TO THE BODY HAVE BEEN EXTENSIVE ENOUGH THAT THE CAR'S EXACT HERITAGE IS NOT IMMEDIATELY EVIDENT

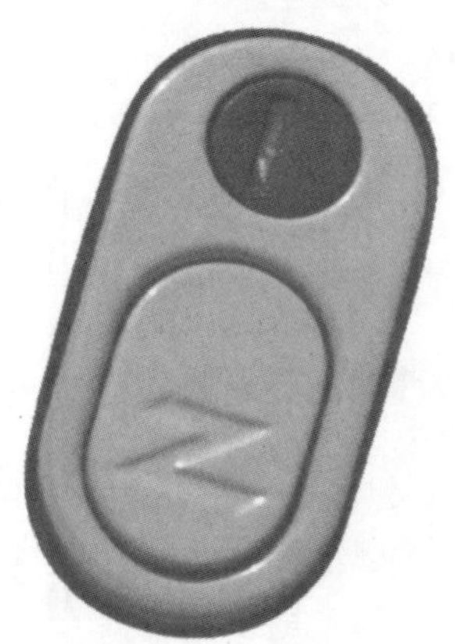

The basic car remains stock, with dimensions identical to the normal 348tb. But the body's lines are slightly more rounded overall. Even the 348tb air conditioning inlet in the chin spoiler has been softened. While most of these changes are quite subtle, in a side-by-side comparison with a normal 348tb a lot of the otherwise unnoticeable differences do jump out. The most dramatic cosmetic changes are the double-bubble roof and huge NACA ducts on either side. The net result is that the car appears to sit more four-square and squat on the tarmac, while the double-bubble roof raises the visual line and imparts this 348 with a more aggressive stance.

If one wants — and can afford — a Ferrari that looks different, then why settle for such subtle changes? Custom bodies are usually radical designs which pave the way for future production cars. And of all the Italian bodybuilders, Zagato's recent creations such as the ES30 "Monster" for Alfa (SCI, April 1991) have been most dramatic, even controversial.

MODERN LIMITS

The real question is whether or not there is room, in this world of legislated crash tests, for a one-off radical conversion. And the answer is straightforward and simple: no. Go more than skin deep and the resulting cost of recertifying a single car for road use is just too high, even if there's an oil well or two in your investment portfolio. In the early sixties, you could buy the best prototype work of Zagato, Ghia, or Pininfarina at the big European auto shows and bring them into the U.S. without being shadowed by Federal officers in cheap gray suits. But that era was closed by a series of increasingly stringent crashworthiness laws dating from the late sixties and early seventies. By the current letter of American law, a one-off car would have to be crashed before it could be delivered to the customer. Clearly, no one wants to ante up funds just to prove Ferrari's ability to make a safe car. Such is the reason limited-production cars like the ES30 aren't sold here.

So, to be law-abiding in the 1990s is to be content with only slightly modified rapture. In this case, it's enough, for Zagato-bodied Ferraris have been few and far between. The total number produced can be counted on the fingers of two hands. In the fifties, a small series of light and powerful Zagato-bodied 250 GTs covered themselves in glory (SCI, September 1990). The last Ferrari Zagato was built in 1974 for Luigi Chinetti on a 4.0 liter 330 GTC chassis. It was characterized by the angularity typical of the seventies and sported a targa top.

FERRARI-ZAGATO MYSTIQUE

Considering the fact that Ferrari is entirely a post-war car, the relation between Zagato and Ferrari is surprisingly long, dating back to the twenties when Ferrari was a major Alfa dealer. Enzo was an avid racer, and he began driving for the official Alfa team in the twenties. In 1934, Alfa withdrew from racing and Scuderia Ferrari became the official racing organization for Alfa. That was the era when Zagato-bodied 1750 and 2.3 Alfa sports cars were invincible.

Elio Zagato, a young nipper of nine at the time, recalls: "Enzo Ferrari and my father became good friends. They had to, really, because Ferrari would spend a lot of time in our factory trying to speed up the preparation of cars for the Mille Miglia. On at least one occasion, workmen were still fitting the bodywork on the transporter en route to the Brescia start!"

Ferrari continued his association with Alfa even when the racing effort was taken back in-house as a result of the German grand prix onslaught. Ferrari remained so intimately linked with Alfa that he was one c those who can reasonably clair authorship of the fabulous Alf Type 158 GP car which, no incidentally, carried a bod crafted by Zagato.

The Zagato brothers' worl dates back to some one-off se dan bodies and a few racing spe cials in the early twenties Thanks to Ugo Zagato's appren ticeship in the aircraft industry they quickly gained a reputa tion for building ultra-light bod ies. It was an era when Alfa along with other expensiv marques such as Isotta-Fraschin and Hispano-Suiza, really mad only chassis. If a client ha enough money for an Alfa chas sis, then he certainly could af ford a custom body. "Produc tion" Alfas carried sedan o cabriolet bodies from Castagna but competing bodies from com panies like Farina or Tourin, were always heavier and there fore slower than the spider bod ies coming from Zagato. A pre war Zagato-bodied car alway had two characteristics: it wa both outstandingly beautiful an very light. Sportsmen chos Zagato over all other bodybuild ers for both reasons.

In the post-war years, th Zagato family was struggling t stay afloat by building produc tion Topolino bodies for Fia But, towards the end of the fif ties, Zagato was once again pro ducing light and exclusive high performance cars such as th Giulietta Sprint Veloce Zagato the DB4 Aston, A6G Maserat and what seems like every Fia that Carlo Abarth could get hi hands on. In fact, it was to gai headroom for the diminutiv Fiat Abarth Zagato 750 coup that the double-bubble roof wa created. Thereafter, Zagato en tered small series productio with coupes for both Alfa an Lancia as well as a 1750 Alfa Zagato replica. In the mid-eight ies, Aston Martin returned t the Zagato fold. The spirit c the legendary DB4 Zagato c the sixties was revived with Ital ı coachbuilt versions of the ıntage, then Volante models. on thereafter followed Alfa ›meo with the ES30, now lled the SZ.

ORTY CHANGES

hich brings us neatly back to e present project. The Ferrari 8tb, like nearly every Ferrari fore it, was styled by ninfarina. The Zagato family s enjoyed excellent relations th Pininfarina ever since ttista Farina bought an Alfa 50 Gran Sport Zagato. So it ıs not Zagato's desire to diınish Pininfarina's work that ove this project, but rather to ve the car special characteriss desired by the customer.

The effort involved no fewer an 40 changes, some subtle, ıers less so. Because neither e engine nor the running gear ve been touched, the car cars Ferrari's mechanical warıty and is certifiable for crashrthiness. It is thus a candite for limited production, and gato plans to offer the same anges to a maximum of 20 ners who wish to convert their 8tbs. The details of each indilually numbered car will be eserved in a register kept by a tary public.

So what's been changed and what measure? Zagato's corrate identity is tightly linked th the double-bubble roofline. this instance, the roof height s not been lowered, but the btle bulges are faithfully folved by the interior headlining provide a tad more room.

The 348tb's pseudo-radiator lle has been replaced by an al, body-colored plate which ries the Ferrari prancing rse logo and a pair of fog hts. At the ends of front bumper, in ıce of the 348's e-piece side and licator lights, there two pairs of clas-, round plastic ıses countersunk o the fiberglass. Of necessity, the 348 headlights remain the same, although the leading edge of the pop-up lid now lies flat with the surrounding panels.

The hood itself has undergone a major transformation: it is now fabricated from aluminum and the original horizontal air vent at the base of the windshield has been exchanged for a pair of eye-catching, NACA-style ducts. The doors are also all aluminum and the original Testarossa-style strakes are now NACA ducts.

The original fender line was an angled fold that ran front to rear. Zagato has provided a smooth surface and rounded the flank. The rear C-pillar has also been delicately softened with a more gradual curve. Long, stalk-mounted side mirrors are a distinctive feature common to both the Testarossa and the 348. The mirrors on the Zagato are small, aerodynamic pods that fit tightly into the angle of the A-pillar. The lower body panels are painted body color rather than the matte black of the standard 348tb.

Toward the rear, the most noticeable change is the three nearly vertical louvers in the rear fender which recall the 250 GTO. The fuel cap cover is competition style with the Zagato "Z" logo on its lock. The "Z" also appears on the door buttons. Just forward of the rear wheel arch a "Zagato Elaborazione" badge replaces the Pininfarina plaque.

The heart of any Ferrari is its engine; to show it off, Zagato's changes include a section of 4.0 mm thick thermal glass set into the aluminum engine cover so that it just frames the cylinder heads of the longitudinally mounted V8 beneath. Adding even more drama, the Ferrari name and logos on the motor have been highlighted with paint. To keep the bay from turning into an oven (with its own window, no less), Zagato has added cut-outs in the engine cover to improve ventilation while the catalyst, located in the rear of the engine bay, is shrouded in heat-absorbing material.

The original 348 back end mimics the Testarossa's, with a rear grille covering a cluster of lights. Zagato has altered this Ferrari's rear to look more like previous Dinos: a flat panel and three round, smoked lights per side which match those up front. Directly above the lights and to the rear of the engine cover, there's an electrically operated spoiler. When it's raised, two simple graphics are visible on its underside: Ferrari on the left, Zagato on the right.

The Ferrari wheels have been replaced by a set of 17-inch,

THE ZAGATO FAMILY HAS ENJOYED EXCELLENT RELATIONS WITH PININFARINA EVER SINCE BATTISTA FARINA BOUGHT AN ALFA 1750 GRAN SPORT ZAGATO. SO IT WAS NOT ZAGATO'S DESIRE TO DIMINISH PININFARINA'S WORK THAT DROVE THIS PROJECT, BUT RATHER TO GIVE THE CAR SPECIAL CHARACTERISTICS DESIRED BY THE CUSTOMER

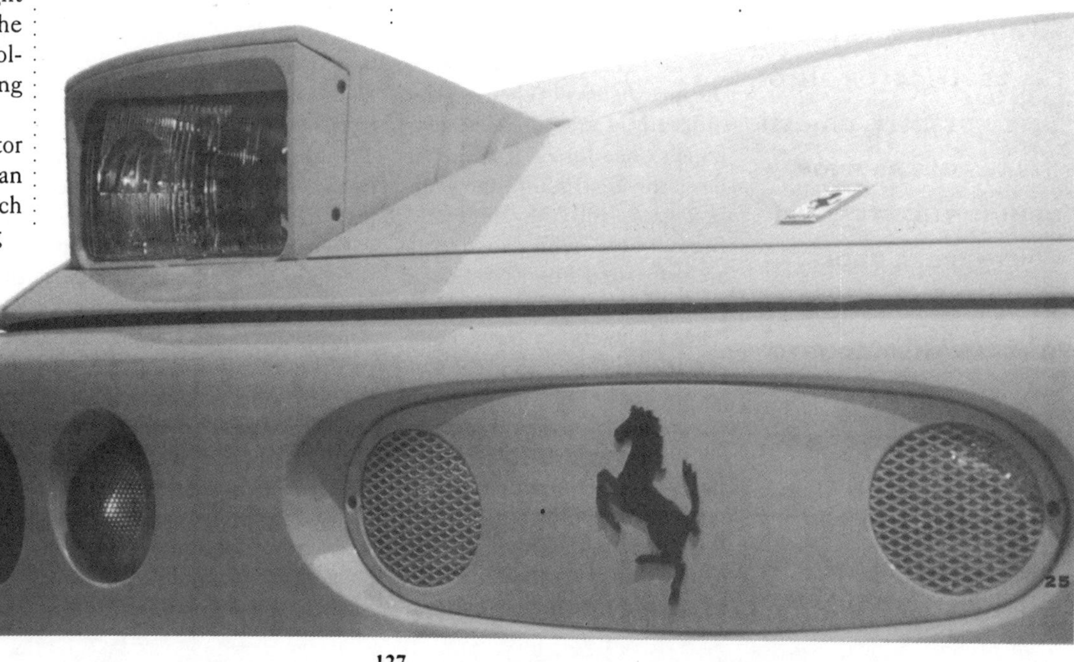

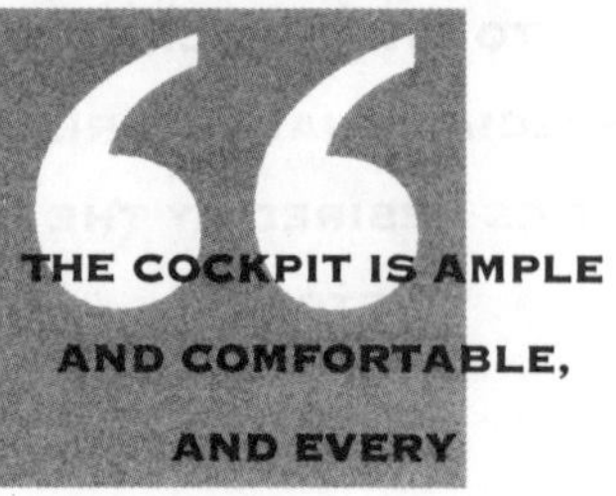

THE COCKPIT IS AMPLE AND COMFORTABLE, AND EVERY CONTROL OF THE 348TB REMAINS IN ITS PLACE. THE SLIM GEAR LEVER EXTENDS ITS COOL-SURFACED ALUMINUM SPHERE TOWARDS THE PALM OF YOUR HAND. PRESS THE LEVER, *CLOK*, INTO THE FIRST GATE. GARISH DIALS GLARE FROM BEHIND THE STEERING WHEEL. A PROD ON THE ACCELERATOR PROMPTS THE ACOUSTIC THRASH OF AN OBEDIENT QUAD-CAM V8, AND YOU'RE UP AND AWAY

Zagato-designed OZs that wear standard 348 tires: 215/50 front, 255/50 rear. The five-spoke design features a polished outer rim with an anthracite gray center highlighted with a yellow center lock.

The remaining external changes hint at what remains to be discovered in the interior: a mini antenna at the rear of the roof and a camera lens peeking rearward from the trailing edge of the roof. This Zagato is equipped with a three-inch color monitor in the central console that switches between a wide-angle rear view and regular TV.

The steering wheel is of Zagato's own design; otherwise, the interior has not changed significantly in structure. The door panels, headliner, and lower dash, the area around the gearbox gate, and the shelf and firewall area behind the seats are all finished in dark gray suede with highlights of stitched-in black Connolly leather. One of the central ribs of the seat pattern is imprinted with a Zagato signature.

The seats themselves reflect Zagato's vast experience pioneering anatomical seats for the Gran Turismo racers of the fifties. The 348's seats are already a great improvement over those of the old 328, but they still don't support the upper torso well. Zagato has inserted extra padding and increased the dimensions of the side bolsters so that the seats' wings latch intimately onto your ribcage, making the car fit like a pair of tight leather pants.

An Alpine CD changer control is placed behind the rear view mirror (there still is a conventional mirror), and right underneath that rests a rocker switch to raise and lower the rear spoiler. The color camera's electronic black box is mounted on the rear parcel shelf between the seats.

On the Road

That's the basic anatomy of this *fuori serie* modern-day special. On the open road, the Ferrari Zagato pops open a few more mouths than a normal 348, probably because its bright paint shocks the sensibilities of the northern Italians. While Ferraris remain rare on the autostradas and byways of the home country, Fly Yellow ones are rarer still.

The cockpit is ample and comfortable, and every control of the 348tb remains in its place. The slim gear lever extends its cool-surfaced aluminum sphere towards the palm of your hand. Press the lever, *clok*, into the first gate. Garish dials glare fro behind the steering wheel. prod on the accelerator promp the acoustic thrash of an obec ent quad-cam V8, and you're and away.

Visibility using the Zagato small side mirrors is excelle although the habit of watching three-inch color monitor rath than an interior mirror is o that grows slowly. The design the prototype's rear spoiler h been troublesome, but wo continues to find the optimu configuration.

There are no surprises in pe formance: the Zagato possess nothing more nor less than t standard 348's cut-and-sla cross-country ability. At spee over the ton the big tires ro the engine gets busy and, d spite a lower drag factor than t 328, there is a fair amount wind noise.

First-time test drives are ha to come by and this one w particularly special because t car is not owned by the facto but by a customer who tru ingly gave his blessing to SC test. This fact, and the nature the traffic around the outski of Milan, meant that there cou be no high-speed verificati that the revised aesthetics ha improved the car's performanc That doesn't take away the fe ing that the Zagato is still a th oughbred with the instincts o racer. And, with Zagato's pan beating, it certainly stands ap from the crowd.

The number of 348tb Ferra to bear the distinctive doub bubble roof will total a mere cars, a pity considering the w that shape flatters Maranell supercar theme. And why su an odd number? Well, fi there's the prototype shov here, then 20 cars destined happy customers. The 22nd w be kept by Zagato. Thanks careful planning, a 23rd wo have to be sacrificed just to s isfy bureaucratic curiosity ov Ferrari's competence to bu crashworthy cars. **SCI**

CO-B10944

Horse Play

AURA, MYTH, MAGIC — MICHAEL BROWNING DISCOVERS THE GLORIES OF THE FERRARI 348ts

The 348 borrows from Ferrari's racing experience only what is appropriate for a road car.
Our test car demonstrated that quality levels at Ferrari are admirably high, with excellent standards of panel fit, paint and hardware.

The prettiest production Ferrari for years, the 348 is a treat for the eyes at any angle. It appears much more purposeful than its longer, leaner 308 and 328 forebears, to create a distinctly squat, stubby appearance.
The 3.4-litre, 32-valve V8 engine develops 221 kW at 7200 rpm and 323 Nm at 4200.

Somewhere between Healesville and Narbethong over Victoria's majestic Great Dividing Range, I nearly changed horses: Maranello's prancing black stallion for Porsche's similar steed on Stuttgart's coat of arms. It's a magnificent twisting, ducking and diving stretch of blacktop on which driving dreams are fulfilled. Cut and thrust, up-change, throttle, heel and toe, down a cog, point and plant. Motoring calisthenics to a rap beat.

I said nearly. For a confirmed Porschephile like myself, Ferrari's glorious 348ts Spyder lifted the curtain on another, perhaps more exciting world — a world of charisma where aura, myth and magic all but obscure the object of desire, and reality is entrusted to the few fortunate enough to sample and savour each marque. Dr Porsche forgive me; it was almost an unfair contest. After an enduring week of malevolent Melbourne winter weather, the June morning was reborn fresh and frosty and the sun laser-beamed through the fog and patchy clouds as I was introduced to Ferrari's latest and potentially most successful production model. The 348 in either closed coupe "tb" (transversale berlinetta) or flip-top targa "ts" (transversale spyder) is the latest in the line of "budget" non-V12 Ferraris which began with the dinky 246 Dino two decades ago.

It's not a Dino. Not in name anyway and certainly not in temperament if you are comparing it to the original 246 model. That was a little gem of a racing car, loosely refined for the road, whereas the 348 borrows from racing only what is appropriate for road use. That's the basic difference. But it's what you'd expect from a company which has evolved from a small enterprise building road cars in order to finance its racing activities, to the appendix of one of the world's largest car makers for whom racing is a group promotional activity. Despite the jibes from purists about the "two" Ferraris — the "real" ones and those with Fiat speedometers — the Dino line through the 308 and more recently the 328 series has been a commercial success, keeping the marque alive and well under Fiat's benevolent wing while other famous marques have been emasculated. The ageing 328 in fact was the most popular model in Ferrari history and its successor is destined for even greater glory.

Around 2500 348s were built in 1990 and the 1991 result will exceed this. Sure that's small beer by Porsche (40,000) standards, but for a still largely handbuilt car in which one can rejoice at seeing handwelded square and round tubing in the engine bay, that's certainly a cracking pace. In the case of the 348, if our test car is the standard measure, quality at Maranello is at a high water mark. Panel fit is excellent, paintwork even and lustrous and the overall quality of the minor fittings and hardware is well above base Fiat spec.

Beautifully proportioned and aesthetically satisfying from any angle, the 348 is the prettiest production "Faz" for many a year. To some that may be a backhanded compliment. "Pretty" is not a word in the "*tifosi*" macho vocabulary, but then again, Julia Roberts wears the same tag very well. Whereas the previous Dinos represented a separate stylistic line to the V12s of the marque, the new 348 shares many themes in common with the about to be replaced big and brutal Testarossa and the rather ephemeral Mondial four seater, yet succeeds better than both.

The side-long strakes which cover the bulging intakes to the side-mounted radiators, and the plastic slats over the rear lights, are direct steals from the Testarossa. But while the TR emerged like a slab-sided aircraft carrier, the same treatment on the 348 is generally smooth and subtle. Perhaps it's bickering, but the dummy blanked-off radiator seems an unnecessary adornment for a car which breathes through its rear flanks. Like the Mondial whose 32-valve V8 it shares, the 348 is true to its lineage in its mid-engined configuration. But where it differs from the 328 is that the dry-sumped 90 degree V8 is longitudinally mounted, while the transmission is accommodated transversely. While the 308 and 328 were long, lean designs, the 348 appears much more purposeful. The wheelbase is 100 mm longer, yet the 348 is actually 25 mm shorter overall than the 328, the reduced overhang and the wheel-at-a-corner look combining with the substantial 130 mm increase in width to create a distinctly squat and stubby shape. By comparison, a Porsche Carrera 2 is some 250 mm narrower . . .

Inside the strictly two seater cockpit, the 348 belies the statistics with a roomy yet cosy feel. This is not a throwback to those early '70s supercar designs, where many an Italian love affair was physically out of bounds, thanks to a low cockpit and a transmission tunnel you could barely see over, let alone leap across. This is a totally liveable workplace, especially for a 170 cm gnome like myself, but completely tolerable for those of much greater stature.

Our Modern MOTOR impression car was the 348ts Spyder version, which is expected to represent about 80 per cent of 348 sales in Australia. But unlike 95 per cent of its local brethren, this Ferrari was finished in mid metallic "Blu Chiaro", not "re-sale red", as Ferrari's favourite colour is affectionately known in the car trade.

The hue is easily explained. It's a personal favourite of Victoria's Ferrari mentor, Lance Dixon, whose Chequered Flag has held the franchise since 1977. And this was his car, just starting to feel willing with 3000 km on the clock.

Step inside the 348 and you are in familiar Ferrari territory. The handsome leather bound wheel, the classic exposed gate for the five speed gearbox and the abundance of leather — pigskin in the case of Mr Dixon's machine — are expected, as is the manual adjustment for the driver's seat. Despite its $226,691 price tag, you don't get electric seat adjustment, nor a radio. The first, Chequered Flag's Stephen Ellett argues, is not a requirement of Ferrari owners; the second is left to the owner's choice, given the bewildering arrange of sound equipment on the market. You also don't get much a view of the instruments. While the steering wheel is laudably adjustable for height, the preferred uppermost position neatly masks out the right-hand mounted rev counter from about 3000 rpm upwards.

This is rather like telling Romeo he can only have Juliet's toes for company, ignoring the bliss which lies beyond all the way up to the 7800 cut-out! Taller drivers could fare even worse! Pardon me Enzo (in absentium) but I thought the rev counter took priority in the cockpit, just marginally ahead of the gearchange and driver. The oil temperature and fuel gauges are also thoughtlessly placed on the central console which means a long distracting glance away from the road. But as the 348's big 95-litre tank provides a touring range of around 700 kilometres, that's mercifully not too often. Fire up the magnificent quad cam V8 and you hear Verdi in full song. It's a veritable symphony of machine-like mechanical noises which is impossible to detect as a V8 on casual notice.

The new north-south engine location doesn't exactly hide the engine's whereabouts, but it certainly reduces the inroads of cockpit noise compared with the model's 328 predecessor with its transverse layout. Idle is a surprisingly high 1000 rpm and combined with a long throttle travel and general high level of mechanical activity, the first impression is that the 348 is fairly low geared. But this is soon dispelled as you ease in the butter-smooth clutch and work your way easily through the five-speed close-ratioed gearbox and explore the seemingly limitless boundaries of this power plant. This is an extraordinary engine by any standards. The on-paper torque of 323 Nm is good but not sensational, yet this 3.4-litre V8 pulls smoothly and effortlessly all the way from idle speed to the upper regions in any of its five forward gears. You can floor the throttle in fourth at 1200 rpm, fifth at 1500 rpm and expect one long surge.

The engine is totally useable with no peaks and valleys. While serious motoring certainly begins around 4000 rpm, only the distinctive hardening of the exhaust note and an orgasmic increase in induction noise bears witness to your more serious intent. This does not feel like a very fast car. It's no match for the near vertical acceleration curve of a 911 Turbo on full boost, yet the performance figures are not all that much slower. More it's the continuous stream of high performance possible which establishes the 348 as an excellent and serious Ferrari. But you don't have to wring its neck to end up smiling. Using just half this engine's rev capacity will still leave more mortal machinery in your wake.

The gearbox is its near-perfect mate. It's one of the tactile elements which makes a Ferrari a Ferrari. First is to the left and back on the traditional polished metal exposed gate and there is strong spring loading in the second/third plane. The change is slightly clunky and sometimes notchy, baulking occasionally on the way down from fourth to third. Put it down to the newness of the car or my lack of familiarity with it — probably the latter, although I have read of other testers complaining of the same problem. Still when you are hurrying, the result is immensely satisfying, the metal to metal contact of lever on gate-end ringing out the changes in time to the howl of valve gear and bellowing exhaust. And to think that racing Ferraris have now abandoned all this in favour of finger-tip electronics . . .

Surging out of Melbourne at 6.30am on unfettered backroads brought an initial impression of ride harshness. Certainly you would expect this with ultra-low profile 17-inch Bridgestone RE 71s front and rear. More familiarity changed this impression. In truth, the suspension works exceptionally well, ironing out bumps and undulations with little protest and rarely a squeak from the removable targa top. The problem really lies with the seats, which are not so much firm, as perhaps firm in the wrong places. Ferrari minder Stephen Ellett said they prove their worth on a really long run and we'll have to take his word for this, although I have my doubts.

But if a firm ride is the price you have to pay for the steering and handling package, it's well worth it. The rack and pinion steering is very direct at 3.2 turns lock to lock and deadly accurate. You can play slalom with every cat's eye lane divider on the freeway and miss ten out of ten.

But rather like the Porsche 911s of old, the steering tends to react quite sharply to road, camber and bumps. The trick is to hold the wheel firmly but not tightly, letting the suspension do the work. Cornering comes into make a wish category, full testimony to the mid-engined configuration and carefully matched componentry. Pick your apex and go for it, letting the 215/50 section front tyres provide instant response with the surety that even harsh power application will be tamed by the 255/45 section rears and the standard 40 per cent lock-up differential.

I drove the 348 hard over familiar roads but never once found its limits. The 348's point of no return is extremely high and best explored on a race track or skid pan. Certainly in wet or dry only the unwary or the uncaring will get into trouble in this Ferrari.

Perhaps the single most confidence-inspiring aspect of the 348 is its huge four-wheel dinner plates, which masquerade as brakes. Internally ventilated and fitted as standard in Australia with ATE anti-locking, the 299 mm diameter front and 311 mm diameter rear discs haul the 1465 kg Ferrari back from any speed you care to name with incredible authority. I can recall no more effective or more perfectly weighted brakes on any road car, with the exception of the new 911 Turbo. In this area, both cars are at the vanguard of a new generation of retardation.

Approach any corner at an impossible speed, squeeze firmly on the centre pedal and the big Bridgestones claw their way into the bitumen, freeing your mind to concentrate on the aesthetics of cornering. This is a wonderful car to travel quickly in, and with the Spyder's removable targa roof, you can share the feeling with nature. Originally scorned by purists, the targa-topped models have become the mainstay of Dino sales over the years. The difference in the previous 308 and 328 models was that the rag-top resulted in the car having a different profile to its tin-topped brethren, whereas in the 348 series, clever design has resulted in an identical silhouette and thus similar aerodynamic values.

Sure there is a small weight premium to cover the extra chassis stiffening, but it's worth it. Body and scuttle shake is remarkably low for a vehicle of this type and even though the "tb" is supposedly 50 per cent stiffer torsionally that the "ts", the Spyder certainly doesn't fall short in this domain. Removing the top is simplicity itself, involving flipping two large front over-centre latches and then sliding the metal-framed top forward to allow the long locating lugs to clear their sockets. It stores neatly and inconspicuously behind the seats, robbing the occupants of around 75 cm of seat travel. This might be a problem for particularly tall drivers, but the rearmost available position still left me ideally seated. At speed, except in topless form the 348 is remarkably silent. Even then there's remarkably little wind buffeting up to 140 km/h — the greatest price you pay is restricted seat travel to accommodate the roof stored neatly behind the seats. Sure there's a little wind noise around the frameless side windows and the targa top creaks at first, but with top on and windows up, it's *conversation normale* at 160 km/h plus.

When it came time to return the Ferrari to its owners, my reactions were mixed. Real Ferraris, legend has it, are uncompromising Sunday cars; Porsches you live with day to day. Did this really fit the legend? There certainly are alternatives. Honda's NSX is the most obvious one at a $60,000 saving. But people who have promised themselves a Ferrari will not settle for a Honda. Then there's the Porsche Carrera 2, a magnificent all-rounder with similar performance to the Ferrari at a Honda price, or the ultimate 911 — the new Turbo — at another $40,000. But Porsche people are generally Porsche people. When you've carried the tablets to the mountain, you believe in the book.

People dream of and yearn for a Ferrari. You can't dismiss that famous bonnet badge as a myth or a wank — it's a very real symbol of a vehicle whose ancestors sowed the very seeds of the sports car philosophy and who have carried it forward with pride to the present day. This is not a sensational car. For that, buy a Lamborghini or a Maserati. Ferrari does not build vehicles like the Countach . . . Ferrari's rarely startle, they evolve. The 348's wheels, for example, are an evolutionary development of the five-spoke design which has appeared on racing and road-going Ferraris for decades, yet subtly refined in this latest model.

The 348 is the everyday Ferrari for the Ferrari aficionado. It is well finished; thanks to the side-mounted radiators it has a real three-suitcase (of sorts) front boot; it's light and easy to drive in any conditions and has sufficient creature comforts to entice milady (or your toy-boy) along for the ride. But when you want it to be a Ferrari it responds like a Ferrari and the 9-5 persona is replaced by "Sunday Special". Forget the myth. Nothing else but a Ferrari really does that . . .

Removing the targa top is a simple affair, allowing the driver closer contact with the outside world while enjoying a car whose rewards come from being driven quickly. With the top on and windows up, conversation is not difficult even above 160 km/h.

AT A GLANCE

ENGINE: 32-valve 3.4 litre V8 engine developing 221 kW (330 hp) at 7200 rpm and 323 Nm at 4200 rpm. Compression ratio 10.4:1

SUSPENSION:
Front: Independent with double wishbones, coil springs and telescopic shock absorbers, anti-roll bar
Rear: Double wishbones, telescopic shock absorbers and anti-roll bar

PERFORMANCE: 0-100 km/h in 5.8 seconds. Maximum speed 273 km/h

FUEL CONSUMPTION: Around 15 litres/100 km/h/

PRICE: $226,691

LIKES: A real Ferrari, yet acceptable and exciting everyday transport. Great looks, wonderful engine, sensational brakes, great road grip

DISLIKES: Over-firm seat, poor instrument layout, steering reaction to road undulations

A FAMILY AFFAIR

"Taking a Ferrari home is an event worthy of a BBC outside broadcast"

There aren't many excuses to be made these days when you go to pick up your new Ferrari. With the latest Mondial t it's all quickly understandable and workable. Except the gearchange. "Don't even try to get second when the oil is cold," said the Mondial's regular keeper as he handed me the keys and a month's pass. "Do like all Ferrari drivers and just use first, third and fifth until everything has warmed up."

The advice is not so much an apology as a statement of fact that you, as someone who has chosen to drive a Ferrari, will accept without question. Ferrari chief Luca di Montezemolo doesn't just accept it but in an interview with this magazine (6 May) has enshrined the stiff gearchange and noisy engine as things without which a Ferrari wouldn't be a Ferrari.

It's meant to be old-fashioned logic that says the more out of reach a car is — for its high price, tough driving challenge and impracticality — the more people lust for it, but there has to be truth in it still, even though Ferrari seems to be covering all options with the Mondial t. Here is a Ferrari it's said you can use every day, one which comes with four seats, power steering and a sunroof, with even a new semi-automatic gearbox on the way. In a telling and necessarily anonymous quote, a Ferrari insider said: "With the Mondial a lot of reasons for owning a Ferrari have disappeared. A woman could drive it."

Aargh! I am the first journalist ever to be lent a Ferrari for a month and a woman could drive it. What is the world coming to?

It's certainly a different world for Ferrari today. The Mondial's body is a mixture of aluminium, steel and glassfibre but not so long ago you would have thought that even this, the tamest prancing horse, was made of gold. You can take delivery of a new one tomorrow, and you will pay less than the cost of a big Mercedes saloon. Offer an attractive cash or part-exchange deal and you will probably get the stereo thrown in, maybe a rounding down of the price. What you couldn't expect to do is sell it the next day for more than you paid for it: the trade would depreciate it by about £20,000 after a year. Immediate delivery, discounts, depreciation? The prancing horse has resisted far longer than most, but for the world's most famous sports car, reality has dawned.

And reality for this particular Mondial t was a most unusual month by Ferrari standards: it was to be used not as a fashion accessory or a weekend plaything, but as a car. Every day. In all weathers. To the corner shop for a bottle of milk and to school to drop off the kids. With four people aboard, and luggage.

Dash is simple, but with that badge it can afford to be

Leather-lined cabin looks great but gets grubby quickly

Given its head on open road, Mondial is swift and sure

I was to do 3000 miles in it in a month, more than the average Ferrari owner does in a year. Snicking the gear lever between third and fourth at 7500rpm on a winding moorland road was going to be important too, but before that the famous exposed gear lever gate had to withstand a fusillade of Jelly Tots being flicked at it from the back seat. Welcome, Ferrari, to the real world.

What do you find when you get your Rosso Corsa pride and joy home for the first time? The neighbours all come out, that's what. The Mondial t was never the greatest looking Ferrari and today, for all its quite excellent packaging and handsomeness, is dated: just look at that improbably long and high front overhang. But, and this is the point, those Pininfarina lines aren't anywhere near prosaic enough for anyone to take it for anything *but* a Ferrari. Taking a Ferrari home for the first time is an event worthy of a BBC outside broadcast.

What you find is a car more than 150mm (6ins) shorter and 150mm narrower than a Granada, and what appears to be an equivalent amount

Pre-breakfast blast across open country roads is fine start to day; handling is as neutral and poised as it appears

lower. The drivetrain is identical to the Ferrari 348's, which means the 3405cc all-alloy V8 pumps out 300bhp at 7200rpm and 234lb ft of torque at 4200rpm. For the Mondial t the quad cam, 32-valve engine, fully catalysed these days and dry-sumped as always, uses the new transverse gearbox, a compact solution that puts the drivetrain in line astern, just aft of the two rear seats. Everything else is just as proper — double wishbones, big vented anti-lock discs, limited slip diff — until you get to the powered rack and pinion steering and three-way adjustable dampers with their suggestion (well-founded) of ride comfort.

The cabin is proper, too: there's a simplicity that borders on the uninteresting, were it not for the elegant detailing (let down only by a horrible plastic air vent) and acreage of exquisite leather. With a layered colour scheme that goes red, cream and black, the cabin takes on the appearance of some exotic gateaux. Take delight, too, in the embossed leather case containing a well-stocked tool kit that includes the aerosol tyre inflater (no spare is carried) and, surely the ultimate in workshop one-upmanship, pliers bearing the prancing horse symbol.

The first drive home was neither intimidating nor hard work. Driving a Mondial for the first time, as Ferrari virgins in the office will testify, is far easier than most people would imagine. For the first and last time in the Mondial, I made the mistake of using my normal, speed-humped route — not a good idea — and the equal mistake of a brisk first gear take-off from the lights only to be passed 100 yards up the road by sundry hatchbacks as I was suddenly confronted by a box full of neutrals. The good news with the gearbox is that you can slot home first and third come what may, and the gap between them isn't large enough to show any serious hole in the V8's torque delivery. But, as the man in the shop said, you need to wait for the oil to thin before relying on second. Really, it's a bit rich in a £67,000 car.

"It sounds like ancient plumbing when you fire it up from cold"

By the time I reached home I had, in true road tester fashion, made a mental list of about a dozen things likely to annoy me over the month. In view of the seat cushion's lowness, flatness and paucity of padding, coupled with a sharing of the footwell with half a Goodyear Eagle, I imagined the most serious of these would be a certain amount of discomfort for someone 196cm (6ft 5ins) tall with most of that between backside and toes: not exactly the proportioning of a typical Italian. Wrong. I was never as comfortable as I am in, say, a long-limbed special like a Porsche 928, but I was never uncomfortable either, even after three hours at the wheel.

How many of these other things on my list would be the same? Does it matter the doors don't open as wide as they could? That the engine sounds like a house with ancient plumbing when you fire it up from cold? That it then goes on to do a first-rate impression of an electric sewing machine in need of a service? That there is precious little provision in the cabin to store anything more than a soft-cover map? That the minor switches are arranged such that all the less important ones are where you can see them and the important ones aft of your left buttock?

Similarly, would it matter that the windscreen bonding was losing its grip, that wires were visible beneath the dashboard, that the rear luggage lid was chafing the bodywork, the seat backrests rocked an inch backwards and forwards and a bit of backing plastic on the dash was coming unstuck? All of this would be a Nissan's undoing, but a Ferrari's too?

The first question on my mind was rather more practical: I did rather want to have a car to give back at the end of our month together. And in pretty much the same as-new condition in which it arrived. Suddenly in the world of park-by-ear, fling-open-the-door merchants and car thieves, the Mondial seemed very vulnerable. Body protection ▶

“Supercar clutches used to have an inspirational feel. The Ferrari’s still does ”

◀ amounts to diddly, anti-theft modifications to a door lock I wouldn’t put money on and a pull-out radio. It wasn’t the radio I was worried about.

Driving a Ferrari changes your habits. You find the car park spaces where none can park next to you; you don’t leave it in unlit corners; I took a taxi to the airport rather than leave it languishing in the long-term and you have rather interrupted dinners when visits to the loo suddenly become far more frequent and are via a route that takes in the restaurant car park.

So what is it that makes this car precious beyond its price and function? Or is it that we are just conditioned to think that way? Certainly most people would agree the mystique of Ferrari, founded on legends of both the man and his racing and fuelled by an irresistible Italian passion, is as enticing as any in the car world. Hype or not, it works: there are no cheap Ferraris, no tatty ones, and none lurking unloved in *Exchange & Mart*. Normally this is treatment reserved for classics no longer made, but to be in such a position and still turn out 4000 or so a year is unique.

And even with the Mondial, Ferrari has been pretty true to itself in making a car that stands for the virtues on which that image has been established. The athletic dynamism, driver involvement and precision are real enough.

It takes a Ferrari to tell you what a well-engineered throttle linkage should feel like after too many cars that don’t know or don’t care. How many power steering systems are this alive between your fingers, how many chassis so talkative (if ultimately so very challenging)? Supercar clutches used to have an inspirational feel. The Ferrari’s still does. You don’t know what well-chosen gear ratios are until you have click-clacked that Bakelite-topped spindly chrome lever between two, three and four at 7500rpm. Perfection. Do this on a diving and twisting country route, heeling and toeing your way down the gearbox as well as charging up in it, and — if you do it well — it’s as invigorating and stimulating an experience as we’re allowed these days.

And of course if you fluff a gearchange or miss an apex you can always go back and do it again: there is always the challenge of being able to drive it better. For all its delicacy a Ferrari is as tough as they come where it counts, and the harder and faster you drive, the more the car starts to work in harmony with itself and the driver. Awkward second gear? Forget it. Double declutch accurately and you can bang the lever up from third into second as fast and cleanly as any. Sound like a sewing machine? Not now it doesn’t. It’s still not the most musical sound but it is all mechanical, and it’s very hard and very loud.

Don’t think that this relies on taking the car beyond its capabilities. To drive a Ferrari well is to drive it fast, yes, but also to use up road space not much greater than the 1810mm, or 71ins, of its width. The incisive turn-in to a corner, the flat body and the prodigious grip coupled to the very clear signals about the direction in which the car’s rearward bulk will want to go if your right foot gets the willies mid corner, these are all sufficient to stay within the wide band in which handling is neutral, perhaps just occasionally nudging into mild understeer. On the scale of effort, say driving at seven to eight-tenths.

We have all seen enough photographs of Ferrari test driver Dario Benuzzi at 10 and 11-tenths, holding his car in wild oversteer slides to see that, in the right hands, such handling is not beyond the car’s capabilities, but it is foolish to pretend it would not be beyond that of many drivers. Certainly mine: playing with the Mondial on a steering pad was enough to confirm that the great feeling of weight transference under throttle lift-off at maximum lateral acceleration is no idle threat. The subsequent oversteer slide is not as quick as a 348’s but it does go on and on and on and you do need a lot of very quickly applied power as well as opposite lock to get it all back in shape. I lost the tail more times than I caught it, for odds I would not care to bet on in public road driving.

What is delightful in the Ferrari is driving it at 70 or 80mph on a country A-road — barely six-tenths motoring — but still take a delight in its responses, precision and poise. At least, with the family aboard, I told myself I could.

Yes, for a small family this car works. Some compromise of front legroom is required but once made there is room for both child seats and children, with effective seatbelts and, the real clincher, plenty of glass so they can see out. Bigger children than Kate, five, and Brooke, two, will find the backrests too meanly padded and upright, something Peter across the road, a 12-year-old used to riding in the much more sumptuously padded and shaped back seat of a 928, was quick to point out. For adults, headroom is the big problem.

After 500 miles to Devon and back in a weekend, no one had any grounds for complaint

With 0-60 in 5.6secs, Mondial can gallop with the best

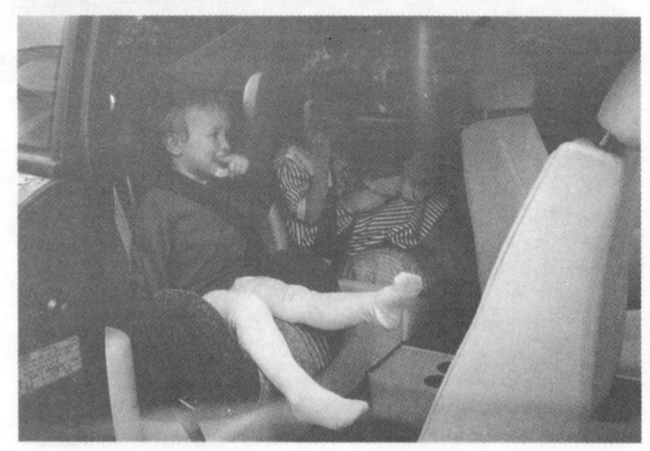

Kids loved sitting in the back, safety seats an easy fit

Every horse needs its exercise; first find an open road

on being too hot, too cramped, too uncomfortable (apart from Cat's-eyes which go off like explosions, the ride anywhere is superb), too unable to hear *Three Billy Goats Gruff* on the stereo (it was close that one, though) or too sad at having to leave a favourite teddy behind. With the front boot able to take a mid-size Samsonite and rear boot a well-shaped oblong, it is surprising what luggage you can squeeze in.

There was only one thing that drove us all wild: the almost complete lack of anywhere in the cabin to put anything. Nevertheless, the kids took to the car instantly and never went off it; they loved the picture of the little horsey so much they insisted on taking the keyring to bed.

Whether the car liked them so much is another matter. Finest Italian leather in light cream is not known for its child resistant properties, although at the end of a month it was the driver's seat that looked the grimy one, with the bolsters creased and the leather tinged blue from jeans. The carpets had started to curl up at the edges and, more worrying, the paint was getting stone chipped and had flaked off some of the external aluminium panels. To use it every day and keep it immaculate would require hours of work a week.

Mondial is still a handsome car, despite age; people automatically know it's a Ferrari

A month showed up no mechanical weaknesses. The only problem was a suddenly very noisy and inconsistent power steering pump that needed topping up. In 3000 miles the V8 got through two litres of Agip SINT 2000, and unleaded fuel at the average rate of 20.4mpg, with a best return of 22.9 and worst of 16.7. Not bad, considering when last we tested the Mondial t we got 154mph flat out and 0-60 in 5.6secs, both of which felt readily achievable.

Other figures don't add up so well. UK Ferraris get used so irregularly that importers Maranello Concessionaires had to devise a special six-monthly service check: leaving a car like this in a garage is as bad as over-using it, they say. Just as well, you say, after looking at the scheduled servicing costs over 37,000 miles: at £6400 plus VAT it's five times more than a Honda NSX would cost at £1169.

What it costs to service a Ferrari

Mileage	Labour	Parts	Total
6000	£295	£158	£453
12,000	£744	£428	£1172
18,000	£282	£173	£455
25,000	£2055	£572	£2627
31,000	£282	£192	£474
37,000	£744	£477	£1221
Total	£4402	£2000	£6402

Note: 25,000 miles/two year service includes recommended cambelt replacement. VAT extra

Another big incentive not to use it is insurance. I asked John Scott and Partners of Farnham, Surrey, what a Mondial t would cost me to insure. Answer: a reasonable £948 a year and no no-claims bonus necessary. Handy for many Ferrari owners who drive a company car and don't have a bonus in their name. The only snag here is that I could cover only 3000 miles a year. That's one good European touring holiday, I reckon. John Scott has 300 Ferraris on its books and most are covered under this limited mileage policy. For unlimited miles, a 38-year-old with maximum four years' no-claims bonus living in a low risk area would have to shell out £1620 a year.

I get the impression no one is that keen to change any of this. To do so would be to bring ordinariness to Ferrari and the inevitable kiss of death, no matter how much more accessible the marque would become. The concessions the Mondial makes are real enough but far from enough. You could use it every day but you would be stupid, or wealthy way beyond the asking price, to try. It's still a special occasion car for high days and holidays, the only Mondial difference being the family can come, too.

No, there aren't many excuses to be made when you pick up a new Ferrari these days. If there were Ferrari wouldn't make them anyway. Ferrari doesn't make excuses, it makes the fastest, most beautiful and exciting sports cars in the world. The only thing I haven't worked out is whether it's Ferrari that's not yet ready for me, or me not yet ready for a Ferrari. Just as well it's only academic... ■

No spare fills front boot

Rear boot ok, tool kit posh

J. ALESI
Marlboro
FIAT
Agip
MAGNETI MARELLI
arexons
PIONEER
SKF
WEBER

348 v F40 v F1

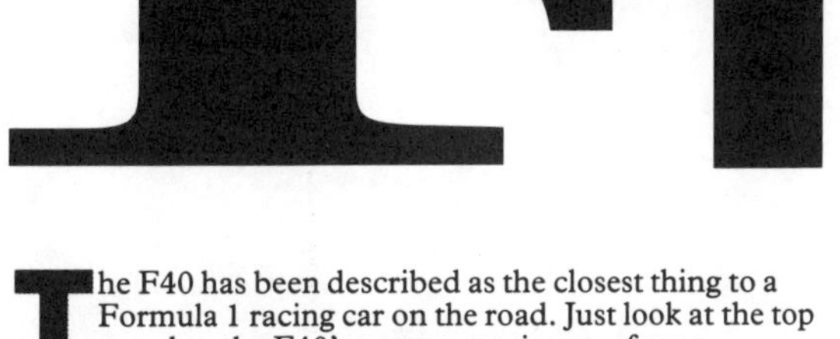

There's a giant leap in performance from Ferrari's 348 to the mighty F40, but how big is the jump from F40 to Formula 1? Drive all three cars round the same circuit, measure their acceleration, cornering speeds and braking power, and you begin to find the answers...

The F40 has been described as the closest thing to a Formula 1 racing car on the road. Just look at the top speed — the F40's proven maximum of over 200mph is every bit as fast as an F1 car goes in the average Grand Prix. And an F40 is stripped out just like a racer inside — no mod cons there. It even sounds every bit as raucous and potent as a racer. But does it really compare?

Michele Alboreto is one man who knows. He drives an F40 in between Grands Prix and he's clocked up more than 15,000 miles in Ferrari's fastest supercar.

'I often and easily exceed 185mph — in Germany, when the traffic is light,' he says. 'The F40 is a real thoroughbred Ferrari. It doesn't just cart you around like other comfortable GTs — it demands concentration and hard work and, in return, offers driving pleasure. It permits extremely high performance in absolute safety; 155mph feels like a doddle.'

So the F40 is comparable with an F1 car, right?

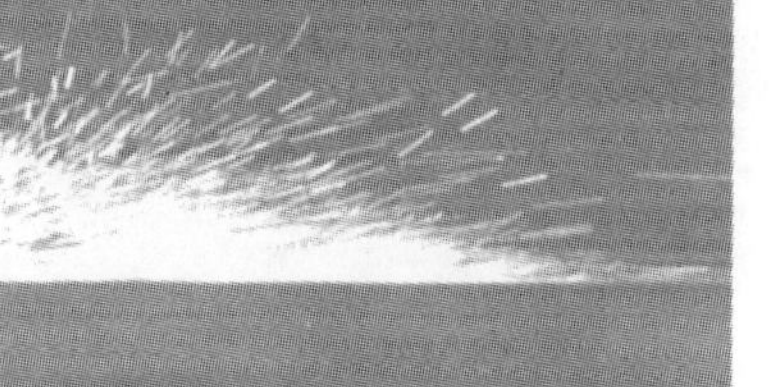

348 v F40 v F1

But what is it about driving an F1 car that's so far removed from driving even the very quickest supercar?

It's not just a matter of speed. Fangio's Alfetta 159 topped 192mph, and that was 41 years ago. The F1 cars of today are little faster, whilst many saloons are capable of running with reasonable safety at up to 150mph, and supercars can do 190mph.

Television pictures of Grands Prix further diminish the sense of pure speed. The cars seem to be cornering on rails, and not even the cameras riding in the cockpits of Senna or Mansell transmit any real sensation, save for the twitching of steering wheel and helmet.

In reality, F1 driving is the greatest automotive challenge that man has ever posed to the laws of physics. The experience has an overwhelming intensity that totally involves both body and mind. Braking on the Monza circuit from 195 to 85mph in less than 100 yards before turning into a funnel-like chicane, producing a deceleration of 3.4g, is not only an act of courage and mathematical precision; it is a technical miracle of which only Formula 1 cars are capable.

In a few seconds the driver has to execute a lightning-fast sequence of actions, dropping down five gears, braking and accelerating at the same time, correcting the line of the car, imposing a trajectory to within 2 inches of the track kerb with the precision of a computer, and all while being shoved forward with a force equating to 600lb — rather as though he were wearing lead overalls.

THREE RED CARS LINE UP on the straight at the Mugello autodrome, the circuit Ferrari uses for testing its cars. Aside from each having its engine mounted longitudinally and just behind the driver's back, they have little in common.

The 348, the car that replaced the legendary 328 in 1990, is the baby of the bunch with a 'mere' 300bhp from its 3.4-litre quad-cam V8. It is a genuine supercar, with a top speed of over 170mph, and it costs £73,591.

The F40 is in another performance league. Launched in 1988, it is the fastest road-going Ferrari of all, its twin-turbocharged 2.9-litre quad-cam

'It's not even close,' says Alboreto. 'It offers the ultimate to a driver who wants to enjoy himself driving, even on his day off, but it's still a road car.

'Jumping from the cabin of the F40 to the cockpit of an F1 machine is like stepping into another world. The limits of the single-seater are completely different, unimaginably so for the ordinary motorist. Not even the driving sensations are in any way comparable.

'When even a well-prepared driver takes the wheel of an F1 for the first time, five laps are enough to leave him gasping. Training is required; a progressive and specific adaptation.'

Even Ayrton Senna had to interrupt his first Grand Prix half-way through because he was suffering from exhaustion; two weeks later, in South Africa, he fainted in the cockpit after the race.

● Baby of the bunch is the 348, with a 'mere' 300bhp from its 3.4-litre V8

● 348 gets a check-over at Mugello

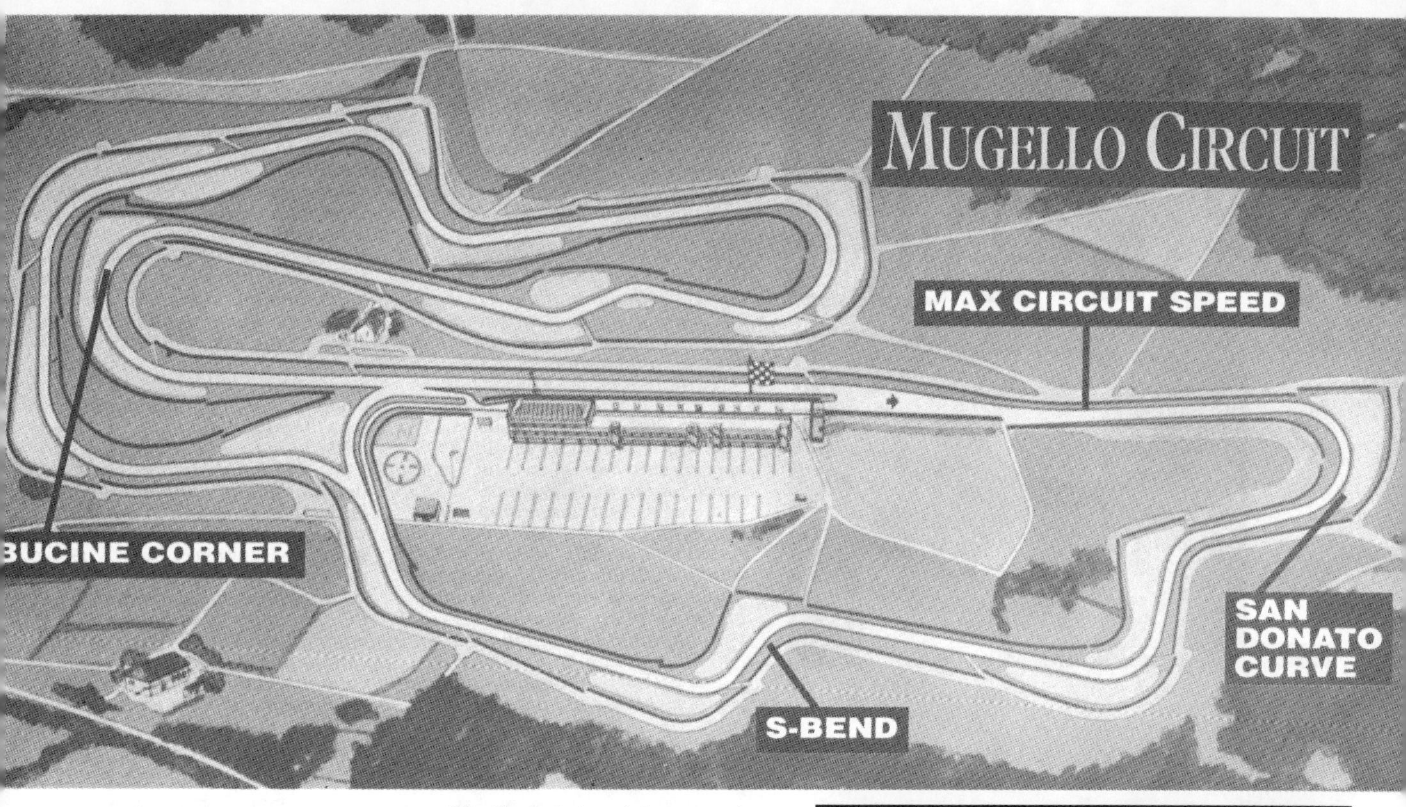

> 'Jumping from the F40 cabin to the cockpit of the Formula 1 car is like stepping into another world'

V8 pumping 478bhp and propelling it to just over 200mph. It used to cost £160,000 but, since Ferrari built so few and has now stopped production altogether, owners tend to name their own price.

The Formula 1 car is last season's 643, but using the latest engine from the F 92 A. This 3.5-litre V12, with four camshafts and five valves per cylinder, produces a reputed 740bhp at a dizzy 14,500rpm. Top speed, at 205mph, is not much quicker than the F40, but the F1 car gets there rather quicker. And the price? Difficult to say, but enough to buy quite a few F40s.

A glance at the power-to-weight ratios is revealing (see specification panel) but what really sets the F1 apart from the 348 and F40 is drag coefficient. The road cars have fairly low cd values — 0.32 for the 348, and 0.34 for the F40 — but the single-seater has to overcome drag factors three times greater (between 0.80 and 1, according to the aerodynamic trim selected for each circuit). Why? Because the F1 uses its ample aerofoil surfaces to provoke negative lift — vertical forces that press the car onto the tarmac and increase as a square of the velocity. At 185mph, the aerodynamic loading can triple the mass of car and driver.

Obviously, this sort of drag limits the top speed of the F1 car — but there's another effect that you'd never

● ***Fastest road-going Ferrari of all — the unmistakable 200mph F40***

Ferrari 348 v F40 v F1

SPECIFICATION

	348	F40	F1
Engine	3405cc V8 normally aspirated	2936cc V8 twin turbo	3497cc V12 normally aspirated
Power	300bhp @ 7200rpm	478bhp @ 7000rpm	740bhp@ 14,500rpm
Weight	3071lb	2722lb	1113lb
Power to weight	219bhp/ton	393bhp/ton	1489bhp/ton

PERFORMANCE

	348	F40	F1
0-62mph (secs)	5.6	4.56	2.5
0-125mph	20.2	11	4.9
Top speed (mph)	173	202	205
Braking (Maxdeceleration)	1.14 g	1.17 g	3.4 g
Roadholding (Max lateral g)	1 g	1.29 g	4.31 g
Timed lap (Mugello circuit 3.26 miles)			
(time)	2min 23.0	2min 9.8	1min 25.5
(av. speed)	82.02mph	90.41mph	137.2mph
Turn-in speed (Bucine corner)	n/a	113.7mph	164.6mph
Max circuit speed	123.7mph	162.8mph	195.7mph
Braking point (for San Donato)	circa 443ft (at 123.7mph)	c 720ft (at 162.8mph)	c 295ft (at 195.7mph)
San Donato curve	52.82mph	58.41mph	74.56mph
Turn-in speed (S-bend)	n/a	69.6mph	96.9mph

experience in a supercar. As Ferrari pilot Jean Alesi observes: 'If you lift your foot off the gas at 185mph, you decelerate as suddenly as if you had slammed on the brakes.' And this is one of the factors that makes F1 cars so difficult to drive on the limit; lift off a fraction and you can compromise your whole race.

OUT ON THE TRACK, the F40 may stay within shouting distance of the F1 on the straights, but when it comes to the twists and turns they might as well be on different planets.

Hit the brakes, and while the F1 creates a g-force of 3.4 under hard braking, the F40 manages 'just' 1.17. Through a fast corner, the F1 is building even more g, this time of the lateral variety. At 155mph it's anchored to the tarmac, resisting a lateral force of 4.31g, four times greater than its mass, while the F40 manages 1.29 through the same bend.

And then you accelerate out of the turn and the F1 picks up from 155 to 195mph with the same thrust that the F40 develops from 0-60!

The track at the Mugello autodrome is 3.25 miles long and rich in gradients. It is also equipped with a timing system at the most significant points, to give us the comparative figures in the table. Unfortunately, a sudden downpour during testing meant some figures could not be obtained for the 348, but there was still enough data to hammer the point home well enough.

Whilst none of the cars was able to achieve its potential maximum speed on the straight, of greatest interest was the huge gulf between the cornering speeds of the F1 and the two road-going Ferraris. Just look at the turn-in speeds for Bucine corner, the F1 entering the bend a full 50mph faster than the F40.

YOU HAVE THE figures, but what is the true significance of those g-forces? As Ayrton Senna says: 'A driver who's very quick over four or five laps isn't going to get anywhere. What's needed is maximum performance over the entire duration of the race, and this is the most difficult target to reach, especially nowadays when over 4g of centrifugal force is recorded in the corners, and we're approaching 5g.' (1g corresponds to the acceleration gravity imposes on a body in free-fall.)

'The F1 picks up from 155 to 195mph with the same thrust that the F40 develops from 0-60!'

● ***F40 used to cost £160,000, but now owners tend to name their own price***

● ***348 generates only a third as much 'g' under braking as Formula 1 car***

Prof Benigno Bartoletti, who's worked closely with Senna in recent years, says: 'The stress involved in Formula 1 driving has a complex web of causes, among which g-forces are particularly damaging.

'Braking deceleration now reaches values of over 3g, and up to 3.5g. They're a little less violent than the maximum centrifugal forces when cornering, but they are repeated more frequently and on all types of track.'

When braking multiplies the weight of the driver's body by three, strange things start to happen to his blood circulation. 'All the muscles contract to resist the thrust,' says Bartoletti, 'and the blood flow is accelerated towards the feet, but slowed in its return to the heart and lungs.

'As the blood remains longer in the extremities it loses more liquids, and this leads to haematosis swelling. The blood becomes more viscous and the return flow through the veins is more difficult. This is happening whilst the brain, at a moment of intense

concentration, requires more nutrients and more oxygen, as do the muscles involved in resisting the forces.

'When not enough oxygen is provided by the blood, the brain tires, its activity slows and it stimulates the heart to increase its pumping efforts. On some less fit drivers we have recorded pulse rates of up to 220 per minute. Insane. . .

'Almost all drivers have swollen feet and legs at the end of a Grand Prix. Many have adopted special Japanese socks with two layers and an air space of about 1mm, which give good grip between foot and shoe, even if they are lightly laced to allow for the swelling.'

Michele Alboreto adds: 'With the cars now having raised noses, we are forced to keep our legs higher, which complicates still further the problems for the circulation. To avoid numbness we have to remember to move our feet a little on the straights.'

One of the most serious effects of lateral g-forces is on the driver's eyes. Bartoletti says: 'These forces often cause distorted vision for just a few moments, but right in the middle of the most demanding part of a corner at 185mph, just when the trajectory has to be perfectly judged and controlled under full acceleration, so as not to come into contact with the kerb.'

SO HOW DOES A DRIVER ADAPT to a modern F1 car?

Michele Alboreto says: 'There's no problem as far as driving technique is concerned; the rhythms of the car are quickly absorbed and become second nature, even for someone coming from Formula 3.

'On the other hand, the extremely high lateral g-forces are worrying because of the physical effects.

'The body is thrust to one side with a force of 300kg. The whole cockpit, as well as the seat, needs to embrace the driver from the thighs to the shoulders. This is why so much care is taken measuring it up — we spend entire days on this job.

'Even so, legs and arms must be left free, and enormous muscular strength is needed to control the posture. This situation is also aggravated by the needle-shaped noses used today, that don't leave room for a heel support, never mind a foot rest to the left of the clutch. The clutch pedal is given a longer travel so that the left foot can rest on it without slipping the clutch.' A driver must still be careful, however. In his first Formula 1 race last season, Schumacher immediately burnt his clutch.

It's the driver's neck that takes the biggest strain — over 4g of centrifugal force means 30-40kg of lateral pull at the head. 'Neck muscles have to be specially developed,' says Alboreto, 'even at the expense of aesthetic matters. Like all the drivers, I follow a physical training programme, and whilst I'm carrying out the exercises I wear a special helmet, weighing 8kg. If you let the physical preparation slide, after just five laps you'll find your head out of the cockpit in the faster curves.'

But these exercises, though important, are not everything. 'Nothing can take the place of the training you get by actually driving the single seater,' says Alboreto. 'There are always muscles that work harder under racing conditions, or in a different way, compared with the simulated exercises.'

THE FIGURES obtained on the test track sketch a fascinating picture of the relative performance of two top-flight supercars versus a Formula 1 racer. The comments of Alboreto and Bartoletti add colour to the sketch. But if you want the full picture there's only one thing for it — get yourself a drive in an F1 car. As you will see, nothing else really comes close! ❍

● With a 3.5-litre V12 producing 740bhp at 14,500rpm, F1 Ferrari simply can't be touched

■ **Test data and pictures supplied by Quattroruote magazine**

Vehicle type: mid-engine, rear-wheel-drive, 2-passenger, 2-door coupe

Price as tested: $119,540 (1992)

Price and option breakdown: base Ferrari 348tb (includes $8140 luxury tax, $2600 gas-guzzler tax, and $1500 freight), $119,540

Major standard accessories: power windows and locks, A/C, tilt steering, rear defroster

Sound system: none

ENGINE

Type	V-8, aluminum block and heads
Bore x stroke	3.35 x 2.95 in, 85.0 x 75.0mm
Displacement	208 cu in, 3405cc
Compression ratio	10.4:1
Engine-control systems	2 Bosch Motronic 2.7 with port fuel injection
Emissions controls	3-way catalytic converter, feedback fuel-air-ratio control
Valve gear	chain- and belt-driven double overhead cams, 4 valves per cylinder
Power (SAE net)	300 bhp @ 7000 rpm
Torque (SAE net)	229 lb-ft @ 4000 rpm

DRIVETRAIN

Transmission	5-speed
Final-drive ratio	3.56:1, limited slip
Transfer-gear ratio	1.09:1

Gear	Ratio	Mph/1000 rpm	Max. test speed
I	3.21	6.0	45 mph (7500 rpm)
II	2.11	9.2	69 mph (7500 rpm)
III	1.46	13.3	99 mph (7500 rpm)
IV	1.09	17.8	133 mph (7500 rpm)
V	0.86	22.5	165 mph (7350 rpm)

DIMENSIONS AND CAPACITIES

Wheelbase	96.5 in
Track, F/R	59.1/62.1 in
Length	166.5 in
Width	74.6 in
Height	46.1 in
Frontal area	19.6 sq ft
Ground clearance	4.9 in
Curb weight	3292 lb
Weight distribution, F/R	42.1/57.9%
Fuel capacity	25.1 gal
Oil capacity	8.5 qt
Water capacity	6.9 qt

CHASSIS/BODY

Type	unit construction with steel-tubing rear subframe
Body material	welded steel stampings and aluminum stampings

INTERIOR

SAE volume, front seat	47 cu ft
luggage space	7 cu ft
Front seats	bucket
Seat adjustments	fore and aft, seatback angle
Restraint systems, front	manual lap belts, motorized floor-mounted shoulder belt
General comfort	poor fair good **excellent**
Fore-and-aft support	poor fair **good** excellent
Lateral support	poor fair good **excellent**

SUSPENSION

F:	ind, unequal-length control arms, coil springs, anti-roll bar
R:	ind, unequal-length control arms, coil springs, anti-roll bar

STEERING

Type	rack-and-pinion
Turns lock-to-lock	3.2
Turning circle curb-to-curb	39.5 ft

BRAKES

F:	11.8 x 1.1-in vented disc
R:	12.0 x 0.9-in vented disc
Power assist	hydraulic with anti-lock control

WHEELS AND TIRES

Wheel size	F: 7.5 x 17 in, R: 9.0 x 17 in
Wheel type	cast aluminum
Tires	Pirelli P700-Z; F: 215/50ZR-17, R: 255/45ZR-17
Test inflation pressures, F/R	35/38 psi

CAR AND DRIVER TEST RESULTS

ACCELERATION	Seconds
Zero to 30 mph	1.9
40 mph	2.7
50 mph	4.3
60 mph	5.5
70 mph	7.2
80 mph	8.9
90 mph	10.8
100 mph	14.1
110 mph	17.2
120 mph	20.8
130 mph	25.1
140 mph	35.1
Street start, 5–60 mph	6.2
Top-gear passing time, 30–50 mph	8.0
50–70 mph	7.4
Standing ¼-mile	14.4 sec @ 101 mph
Top speed	165 mph

BRAKING

70–0 mph @ impending lockup	177 ft
Fade	**none** moderate heavy

HANDLING

Roadholding, 300-ft-dia skidpad	0.92 g
Understeer	**minimal** moderate excessive

COAST-DOWN MEASUREMENTS

Road horsepower @ 30 mph	7 hp
50 mph	16 hp
70 mph	33 hp

FUEL ECONOMY

EPA city driving	**13 mpg**
EPA highway driving	18 mpg
C/D observed fuel economy	**21 mpg**

INTERIOR SOUND LEVEL

Idle	65 dBA
Full-throttle acceleration	86 dBA
70-mph cruising	77 dBA
70-mph coasting	77 dBA

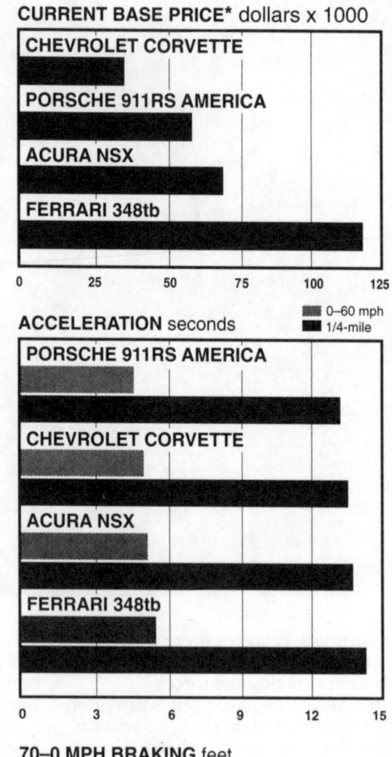

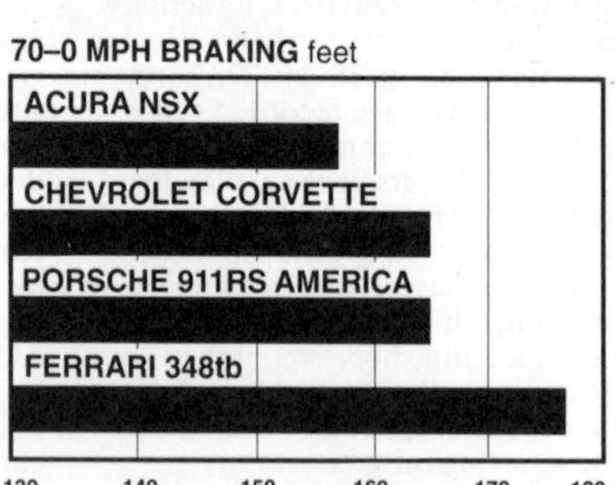

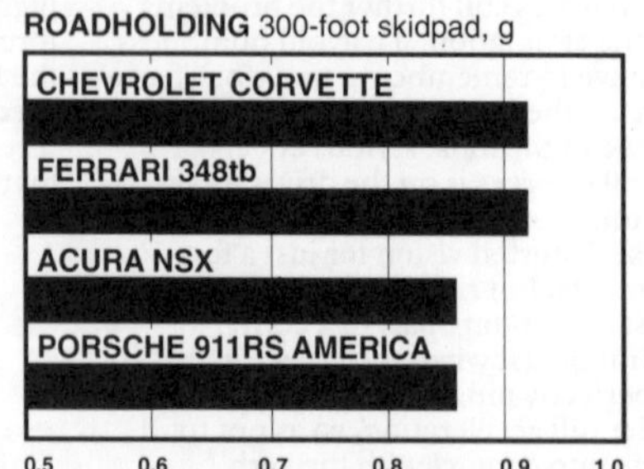

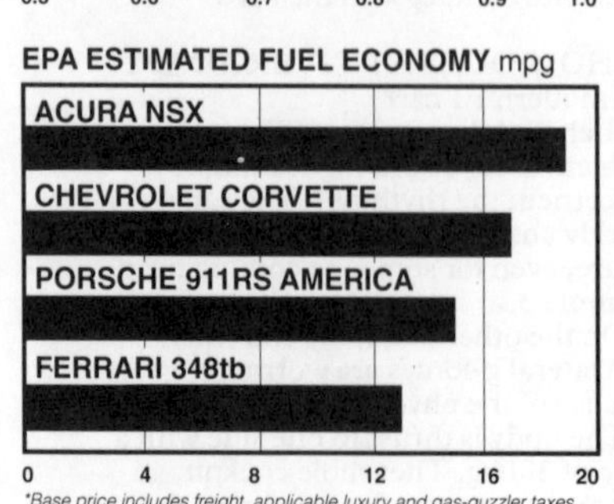

*Base price includes freight, applicable luxury and gas-guzzler taxes, and any performance options.

Ferrari 348tb

This prancing horse has teeth marks on its heels.

BY DON SCHROEDER

Before you gasp at the Ferrari 348tb's sticker price, consider all the perks that inevitably come with owning one of Maranello's legendary prancing horses. Such as how other drivers always let you merge into traffic. Or how valets always park your car in front. The benefits are as endless as they are eccentric; Emilio Anchisi, a past president of Ferrari of North America, firmly believed that the purchase of a Ferrari increased not only your charm and self-esteem, but also the size of, well, your Jockeys.

With that in mind, let's try not to choke on the price: all told, with luxury tax, it's $119,540. *Ooof!* After you swallow that nitroglycerin tab, consider this: even though the price doesn't include a sound system or cruise control, this is still the cheapest Ferrari. Since their 1990 model-year introduction, the 348tb (coupe) and 348ts (targa) have been Ferrari's entry-level models, replacing the 328.

As one expects from top-rank sports cars, Ferraris have always had plenty of sheer cornering grip, but we've noticed some disconcerting handling traits with previous 348s. For one, the car tended to wander, particularly at high speeds, which surely irritated buyers who spent a wad for the express purpose of driving fast. Ferrari conceded there was room for improvement with the 348. Last year, the company swapped shocks and springs, modified the rear-suspension geometry, and made camber and toe changes, all intended to improve the ride and handling at all speeds.

As with past 348s, the steering is so sensitive and telegraphic that it seems at times almost hard-wired to the driver's cerebellum. Cruise over broken pavement and the steering wheel kicks rapid-fire, responding to every crack, crevice, and pavement chunk. Turn the wheel and the car vectors into a different direction almost instantly.

But even with the suspension changes, this superb response still has its downside. Take your right hand off the wheel to shift through a fast sweeper and the Ferrari feints and dodges unless your left hand firmly stays the course. That's not always easy: the 348 has no power assist, and there's strong on-center bias, too. Top-speed behavior is

PHOTOGRAPHY BY AARON KILEY

the Verdict:

Highs: Shapely form, symphonic engine, big-league social statement.

Lows: Medieval shifter, moody handling, incongruous bottom line.

The Verdict: Will run with the big dogs. At twice their price.

hardly improved. Above 150 mph, the wandering on the high-speed oval verged on scary. Of course, few people drive on public roads at this clip. Yet Corvettes, 928GTSs, NSXs, and Vipers can tackle their similar top speeds with nary a misstep. What's the deal with this legendary marque?

But let's pause for a moment and ponder those Ferrari benefits at slower speeds. Whether you're into street challenges or not, this car is going to attract a lot of them. Cruise along any multi-lane highway and Mustangs, Camaros, and sundry other performance cars will buzz around you like flies. At stoplights, you could not attract more attention if you were perched naked atop a unicycle.

You'll win most of those races, thanks to Ferrari's midship-mounted DOHC 3.4-liter aluminum V-8, which is slightly more powerful this year thanks to several engine-control refinements. Dig your spurs into it and it will push the 348tb to 60 mph in 5.5 seconds and through the quarter-mile in 14.4 seconds at 101 mph. Though it's always flexible, the V-8 gets rather feisty above 4000 rpm—and gets plain wonderful near its 7500-rpm redline. The sonic accompaniment is pure Ferrari: each gear rises to one big symphonic crescendo of unmuffled intake air, whirring cam chains and belts, and near-open exhaust. Idling around, it's different still, a combination of Porsche 968 and Oster blender. "This engine sounds very strange," said one passenger, climbing into the seat. This distinctive soundtrack—hardly a bad thing—is gratis with all Ferraris.

Ferrari's transmission, with its trademark gated shifter, is becoming less tolerable as the years go by. You pay a price in speed: moving the lever around in the pattern (especially on 1-2 and 3-4 shifts) requires time and deliberate effort. The shifter effort is but a foretaste for other controls on the 348tb. At the risk of sounding sexist, we wonder how many wealthy women would consider purchasing this car. Certain buttons and latches, like the outside door button and the front-trunk lever under the dashboard, require extraordinary one-finger strength. Ferrari said these items were out of adjustment on our test car, but with the prevalence of stiff controls, from the inside door handles to the steering to the shifter, it's hard not to conclude that they were simply designed that way.

But, ah, those Ferrari perks. Picture this scene in a downtown Detroit restaurant: After sitting down, a friend introduces one of us to the table. Small talk. Friend casually mentions "we came down in his Ferrari." Momentary silence, gaping stares from around the table. (Is this fun, or what?)

Visually, the 348tb makes a fitting statement. Its Coke-bottle shape, side intake strakes, and familiar front and rear end caps turn the 348tb into a pint-sized 512TR. It's certainly a form that quickly distinguishes Ferraris from other sports cars. The interior is equally chic, with crisp-smelling black leather covering most of the insides: the seats, console, dashboard, door panels, and even parts of the running-board trim are cowskin-clad. The insides are as angular and distinctive as an Armani suit. Even the fluorescent-orange space-age numerals on the gauges lend excitement to the instrumentation.

COUNTERPOINT

•I expect Ferraris to be quirky. On a cross-country dash in a 512TR, the motorized belt hit me in the head 23 times before I stopped counting. However, the 348tb's motorized mice *surpass* quirky. They are dangerous. Warn your passenger before twisting the ignition key. Otherwise, the belt can grab your date's right leg, peel her white high-heel half off her foot, and wedge both items—passenger's silk-lined leg and shoe—between the seat and the door. This damages her shoe. This damages the Ferrari's black leather. This damages your prospect for another date. *—John Phillips*

If you must be seen in a Ferrari but can't quite swing a 512TR or F40, go ahead and buy the 348tb—it'll serve your valet-parking purposes nicely. But if you seek the pleasure of driving a fine exotic car fast and hard, keep looking. While its price is exotic, the 348 isn't exotic-car fast, it doesn't handle exotic-car well, the seats are uncomfortable, and the build quality is quite low. The bobbed-tail styling doesn't even measure up to more contemporary exotics. There are cars that cost well over a hundred grand and offer tangible value. This just isn't one of them. *—Frank Markus*

Like the reflexively exhibitionist Madonna, this Ferrari lures onlookers with its flashy red paint, bulging tires, and muscular bodywork. The attraction isn't totally superficial—the 348's screaming engine, grippy suspension, and low-slung ambiance make for as much entertainment as does a Madonna concert. But despite the Ferrari's ability to shed rubber as readily as the venal singer sheds clothes, it can't compete with an NSX any more than Madonna's singing can match Linda Ronstadt's. The 348 and Madonna are both hot dates, but I wouldn't want to live with either one. *—Csaba Csere*

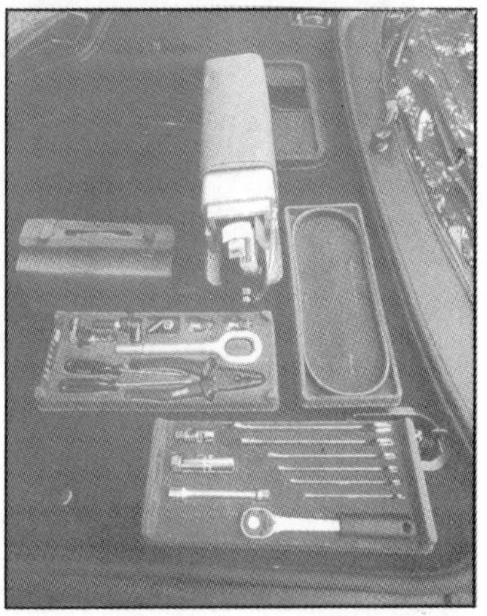

The interior is comfortable, to a point. The 348tb comes with automatic climate-controlled air conditioning and a straightforward control panel. The system worked flawlessly during our test. The seats are starkly firm, but they support their occupants in the right places. The driving position is less than optimal, though, as the front wheelwells intrude on leg space, forcing legs toward the center tunnel. The height-adjustable steering wheel does much to ameliorate the monkey-like Italian driving position, although it sometimes blocks the major gauges.

We wish all the components on our test car were as stout as the 348tb's remarkable flex-free body. The lower door-frame power shoulder belts took an occasional vacation. The driver's seatback latch failed, and the carpeting was pulling away from the passenger door frame. And after our first test of hard acceleration at the track, it felt as if some prankster had poured a bucket of sand down the shifter box. Two transmission shift-rod bushings had worked loose, and they required replacement at the dealership.

But forget those lows for a moment and take the wheel again. On smooth pavement and at under 100 mph, the 348tb offers confident handling, with some of the most controllable and playful oversteer offered on a production car—and that's a definite improvement over the earlier 348s. The ABS-equipped brakes, although they could be more powerful, have excellent pedal feel. And then there's that engine. One charge to the redline makes any test driver consider a second mortgage.

The 348tb remains a worthy descendant of a long line of vibrant Ferraris. But if this marque is to stay at the head of the pack, it still has work left to do on its entry-level car. The 348tb has some faster and more sophisticated contenders nipping at—perhaps even taking chunks out of—its heels, and at far lower prices. Of course, there's a big difference: cars like the Viper, the NSX, and the ZR-1 can easily keep up with the 348, but none of them has a first name spelled F-e-r-r-a-r-i. •

348 SPIDER

Feel Good Factor

The new 348 Spider is destined to become the most popular model in the Ferrari range all over the world.

The United States had first crack at the 348 Spider. Fair enough, really, when America accounts for a third of Ferrari's sales. The first two-seat Ferrari convertible since the 1970s Daytona Spyder could not have asked for a better start in life. Back in February, the first one was auctioned for charity by Sharon Stone in Los Angeles – we would buy any car, new or used, from her – and the fabulously expensive Rodeo Drive was given over to a Ferrari *concours d'elegance* which included some of the factory's Formula One cars.

The Spider went on sale in Italy earlier this Summer and not long after came our first chance to drive it – at Ferrari's Mugello racing circuit near Florence.

Mugello always was a terrific track – hilly, twisty and challenging – but now it is a Grand

Prix venue-in-waiting. Ferrari has remodelled and up-rated the 3.3-mile circuit to provide a longer, faster and more modern test facility than the track alongside the racing department at Fiorano, close to the factory. The Ferrari driving school (£1,300 for a two-day course, Ferraris provided) operates there under the direction of former F1 driver Andrea de Adamich, and one day, Ferrari hopes its massive investment will be fully justified by securing either the Italian or San Marino Grand Prix.

That's enough about the venue. It is important to the 348 Spider story only inasmuch as most full convertibles, without strengthening roll-over hoops or Targa tops, get a shudder through their bodywork on uneven surfaces. The 348 Spider had no such tremors, but then the Mugello track has scarcely a bump in its grippy new Tarmac. On the other hand, cornering hard and running up on the vibration kerbing (intentionally – honest...) put quite a loading on the car which ought to have shown up any lack of stiffness in the body. It did not.

Ferrari says that the Spider's torsional rigidity matches that of the 348tb coupe and targa ts. Even more impressive is its claim that the beefing-up of the chassis side members and the windscreen pillars and frame has increased the weight compared with the original models by only 22lb.

Mechanically, there are a number of detail revisions prompted by the new model but now applied to all members of the 348 range. The 3.4-litre V8 engine has a new exhaust system and stronger valve springs which are better able to cope with regular running to 8,000rpm. A new type of timing belt is 'lifed' for three years. A lower final drive ratio allows quicker acceleration. Higher-geared steering, plus slightly softer springs all round and new Bilstein shock absorbers at the rear improve handling and comfort. Special new versions of Pirelli P-Zero and Bridgestone Expedia tyres have been developed for the 348; the engineers say that the characteristics of both are very similar.

Pininfarina's design for the 348 body needed more changes than one might expect to open up completely. The Spider was not anticipated at the time of the tb and ts launch – the convertible's development started in 1990.

The Mondial Spider had already shown that to produce a mid-engined convertible with nicely-balanced lines required some visual trickery. Hence the canvas 'flying buttresses' at the rear, which avoid a thrown-forward look with the hood up but still allow access to the engine compartment directly behind the cockpit. Folded down and enclosed below its soft vinyl cover, the hood is unobtrusive.

The Spider's engine air intakes had to be relocated behind the side strakes (the tb and ts take air in alongside the quarter windows), the rear deck lid is new and the front grille has been altered slightly. Where the sills of the other versions are black, the Spider has its lower body sides in body colour. Most significant of a number of changes to the interior are the seats, which are designed to incorporate the safety belt mechanism.

On the track, top down and windows up, the Spider is a delight. Even at more than 100mph, wind buffeting is slight. With the side windows down there is more disturbance, but at lower speeds it is more fun. And being out there in the open air allows the driver to appreciate fully that crisp V8 sound.

The car feels taut, not only in apparent rigidity of its body, but also in its precise and predictable handling. The 348 was criticised by some at the time of its introduction as being unstable at high speed and tricky near its cornering limit. The revised suspension and improved tyres seem to have brought a useful improvement.

3.4-litre V8 is now stronger thanks to new valve springs and new type of timing belt

Spider is well-sorted and a delight to drive. Scuttle shake and buffeting are minimal, handling precise

On a sunny Summer's day there was no encouragement to drive the Spider with the hood up. It is a simple canvas roof with a flexible plastic rear window but the factory testers tell us that it works well – no significant wind noise below 100mph and less than two inches of 'ballooning' at the 170mph top speed.

In the interests of research, we did raise and lower the roof. A turn of a single handle at the centre of the windscreen rail is all that is required for tension or release. The hood is light and easy to fold back, though the resulting roll on the rear deck looks messy and demands covering with the fiddly tonneau.

Any 348 is a joy to drive. The smallest Ferrari is the kind of car you feel good to be seen with. But it is not your everyday transport. In Britain, the average Ferrari covers less than 3,000 miles a year. For high days and holidays, an open 348 is even nicer than a coupe or targa-top. We can see why Maranello Concessionaires expects the Spider to account for two-thirds of future sales. **– Ray Hutton**

PRICE

To be announced later this month; about £80,000

COSTS

Insurance group 20

DRIVETRAIN

300 bhp, 3,405cc, V8, mid-mounted longitudinally, driving the rear wheels through transverse-mounted five-speed manual gearbox. Manufacturer's figures 0-62mph 5.6sec. Top speed 170mph

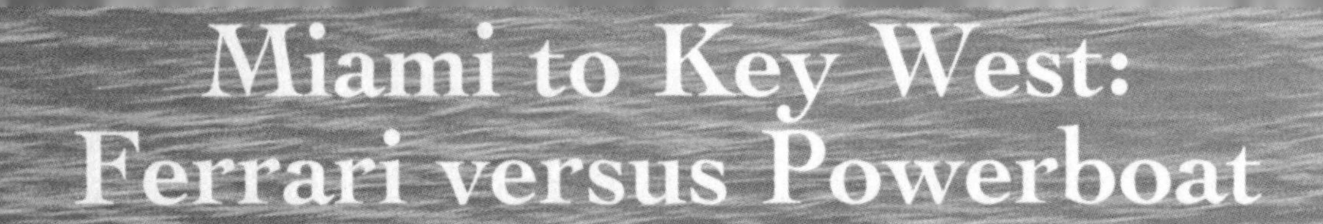

Miami to Key West: Ferrari versus Powerboat

WON if by land, TWO if by sea

by Michael Brockman

PHOTOGRAPHY BY PHIL MASZAK AND MARK SPENCER

It was the sort of challenge that ultimately would prove nothing. No record books would be rewritten to reflect the outcome, and life as we know it would proceed apace, whatever the result. These facts failed, however to dull the challenge. The gauntlet had been thrown, and we were damn well going to pick it up.

As such things are wont to do, this escapade just sort of happened. An acquaintance of *Motor Trend*'s editor called out of the blue with a proposition. *Powerboat* magazine Editor Lisa Nordskog challenged *Motor Trend* to a run, car versus boat, from Miami to Key West just to see who'd get there first. As our staff's native Floridian and prime

knee-jerk gauntlet picker-upper, I was dispatched to work out the details. That Nordskog is an intelligent, persuasive, and attractive young woman had nothing to do with my decision to accept this critically important assignment.

The *Powerboat* people had a new 42-foot off-shore boat ready for sea trial, and they wanted to make a straight run between the two points—a regular route on the boat-racing circuit. But I felt that would give them an advantage because the Keys are famous for two-lane roads and heavy traffic. Elaborate negotiations ensued.

My choice was a course across Florida from Stuart on the east coast through the St. Lucie Canal and Lake Okeechobee to Fort Myers on the west coast. It would give a good car a chance to run, but the boaters whined of locks and no-wake zones, so we finally agreed on a heads-up run from Miami's Biscayne Bay Marriott to the docks of the Galleon Resort in Key West. Yes, their plan right from the start.

At this point, I knew only that along with Nordskog, my challenger would be a gentleman named Fred who was just taking delivery of the boat, called *Off Duty,* which would be shipped to Detroit in 10 days.

The *MT* staff bantered about which car

The Ferrari's interior, is remarkably spartan especially compared to that of the Off Duty. We never did locate the switch for the bilge pump.

would be appropriate. Because we were short on time, we called Shelton Sports Cars for help. Tom and Steve Shelton are both top-notch racers who not only understood this kind of madness, but just happen to own a Jaguar/Porsche/Land Rover dealership in Naples and a Ferrari store in Ft. Lauderdale. They came to our rescue and supplied us with a beautiful, white Ferrari 348ts.

"Fred" turned out to be Fred Kiekhaefer—engineer, world-class powerboat pilot, accomplished businessman, son of Mercury Marine founder Carl Kiekhaefer, *and* president of Mercury Marine Hi-Performance Group. I felt like the gambler who called the Sundance Kid a cheat.

After flying to Miami and picking up the Ferrari, I drove to Biscayne Bay to meet Fred and have a look at the boat. What a stunner: a custom-designed McManus made specifically for Kiekhaefer. As I suspected, this man, who grew up with powerboat racing and who heads probably the most powerful force in the sport today, knows what he wants and how to get it. Words can hardly describe the *Off Duty*.

It's more like a 42-foot piece of floating art than a boat. Most impressive is the color coordination and compulsive attention to detail throughout. From the single starboard bow rail to the aluminum diamond plate in the bilge, everything is anodized, painted, or upholstered to complement the overall design theme. Under the massive hydraulically controlled sundeck/hatch lie a pair of 800SC engines from Mercury Hi-Performance. In fact, all the running gear is either genuine Mer-Cruiser or Quicksilver. You guessed it; SC stands for supercharged and 800 means horsepower. So they're packing 1600 horses as compared to the 312 in our Ferrari.

The engines are connected to a pair of number 6 racing drives fitted with four-bladed 17x27-inch stainless steel props. All this was foreign to me, but the gigantic canary-yellow stern drives with matching trim tabs did look impressive. What was more amazing was the *Off Duty* on the water. I took a short trip behind the helm, with Fred manning the throttles. The sea had a light chop, but the boat flew over the water like a bird through the air, and when Fred indicated a turn at 70-mph plus, I tentatively eased into it. He said to give her more. (I figured it was his boat, so why not?) She (all boats are female, they tell me) took all I could give her at over 70. My thoughts fell back to the impending contest: I knew I was in trouble.

Race day dawned warm, wet, and dreary. I was raring to go, but the boaters were concerned not only with thunderstorms (rightly so), but with the water conditions outside Government Cut, the waterway leading from Biscayne Bay to the Atlantic. They worried that if the swells were over four to six feet, they'd have to run inside the Keys and be mindful of sandbars and no-wake zones. I worried that if the water were calm, they could run outside at near the *Off Duty*'s radar-clocked top speed of 73 mph all the way to Key West. I figured the wet weather would work in my favor. Pavement conditions appeared to be fine, with no unusual swells or whitecaps. I could make just as good a time in the wet as in the dry, maybe even better because the rain would keep away pesky tourists, who normally clog up the Keys.

By 11 a.m, though still overcast, the rain had stopped. We were off. The Coast Guard had told us that the rain had actually helped conditions and that the seas were calm. So as soon as the *Off Duty* cleared the no-wake zone (about a half mile out), it would be flat out to Key West—except for a fuel stop.

The *Off Duty* holds 320 gallons of high-octane fuel in three separate tanks, enough to make it about two thirds of the way down to Key West. That figures to be

This isn't your typical powerboat; all switch functions normally identified with words have been personally modified by Kiekhaefer. He designed every icon for the control panel, also known as K-toons.

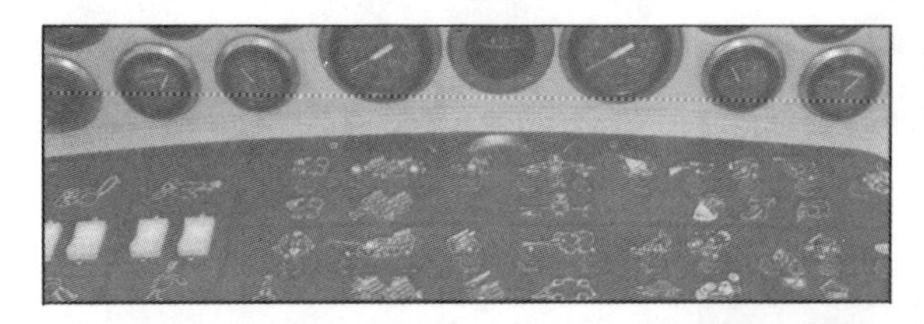

The engine bay, under the hydraulic sun deck, is spotless and laid out with service in mind. Note the spare props mounted to the right.

about 150 gallons per hour for 0.48 mpg, which makes the Ferrari's 18.3 mpg sound pretty thrifty.

It sounds even more thrifty when you compare price tags: We picked, at just under $100,000, what we considered a pricey automobile. We never did get a firm figure of the boat's value, but with two engines at $55,750 each, two stern drives at $25,000 each, a $30,000 paint job, a carbon-Kevlar hull, and hand-stitched upholstery, you can see why the *Powerboat* people had said price was no object in this little duel.

As the *Off Duty* thundered east, I headed west across Miami toward the Florida turnpike. As expected, the trip to Key Largo was quick and uneventful, but there the traffic started—sometimes, 5 to 10 miles of two-lane blacktop at 35 mph. So at any opportunity to squirt ahead (within the bounds of propriety, of course), I took advantage of it. Things were going smoothly until about Islamorada, where a sheriff passed me going the opposite direction. When he saw the Ferrari, he flicked on his "blues" and flipped a U-turn. This I didn't understand, as I was stuck in traffic going 30 mph. It turns out, as luck would have it, in one of my windows of opportunity, I had passed an undercover agent who was clearly impressed with the 348's remarkable acceleration.

After I explained that I wasn't driving recklessly, and after the nice officers discussed little-known vagaries of the penal code for about 10 very long minutes, they welcomed me to the Keys, presented me with a warning, and promised me that if I didn't keep it under 55 mph all the way to Key West, I'd see them again.

After a few thank you prayers and a deep breath or two, it was back to business. As I started across the seven-mile bridge two thirds of the way down the Keys, I glanced east and locked onto an

ROAD TEST

FERRARI 348tb SERIE SPECIALE: The Flavor, the Excitement, the Price

It's a Ferrari. To the faithful, that's ample justification to crave this limited-edition, performance-enhanced 348. If you're not a *tifoso* (in Italian, "to have firm plans for a large inheritance"), nothing we'll say will convert you...but here's a shot: Imagine it's just before sunrise and you're piloting this blood-red beast down a deserted canyon road. The song of the four-cam V-8 flushes a covey of quail. Seventeen-inch Pirellis moan softly while your neck muscles strain against their enormous cornering force on this asphalt roller-coaster. With the massive torque available, you steer almost as much with your right foot and hurtle down a tunnel of live oaks. Now see yourself burbling slowly up to your college reunion to be greeted eagerly by the same sorority girls who spent every weekend evening 20 years ago caring for sick aunts, but who now discover exciting aspects of your personality.

With sales lagging, Ferrari aspired to add appeal to the 348 for can't-drive-155 Americans. First, the roof was removed to create the Spider (Italian for "designed to be driven with the top down and a blonde in the right seat"); it's the

Visual changes include body-color lower trim and engine cover, and revised grille and taillights. New ground-dragging airdam features a replaceable lip. Against the rev limiter in fifth gear, top speed is 163 mph. Seat-mounted buckles are sure to snag sequins from your girlfriend's gown; not a problem unless your wife's gown lacks similar sequins.

ominous vision. About a mile off shore, running at about 70 mph and throwing a roostertail two stories high was what could only be the *Off Duty*. I picked up my mobile phone and dialed. Fred has one of those aircraft-type digital noise-canceling systems onboard, which allows you to listen to music, talk to crew members, or use the telephone via a voice-activated system.

Fred answered; it was them. A dead heat two thirds of the way through—unbelievable. As we were on the phone, Fred made a hard turn west under the bridge to make a fuel stop at Faro Blanco Marina for the only high-octane available between Key Largo and Key West. The good guy in me said I should turn around and greet them at the dock; the racer in me said, "See you in Key West," and hung up. I should've called collect.

As anticipated, the rest of the run to Key West city limits was fast and delightfully uneventful. But the five miles through the city and old town took almost 30 nail-biting minutes, and I was convinced I'd hear the rumble of the twin big blocks idling in the slip as I slewed in to the Galleon resort. After 3 hours and 7 minutes, I turned off the Ferrari at the water's edge, and to my surprise, no Nordskog nor Kiekhaefer. Victory.

Twenty-nine minutes later, the *Off Duty* carved around Fort Zachary Taylor and entered City Marina. Fred and Lisa were wet and tired; I was cool and collected. The boat had burned 500-plus gallons of fuel; the Ferrari had used 10. They downplayed their defeat by raving about how well the McManus handled the sea and what a sweetheart the boat had been on its maiden voyage. I thought what a pleasure driving the Ferrari had been, wishing only for more curves and a change in terrain. And I couldn't help thinking of Paul Revere: Won if by land, two if by sea. **MT**

---

first two-seat Ferrari convertible since '74. Next, the company created a performance-oriented version, the Serie Speciale, available in fixed roof—tb—and removable roof—ts—versions. Over last year's 348, both the Spider and the Serie Speciale have wider rear track to improve cornering stability, a new exhaust system for a bit more horsepower (and more important, better exhaust melodies), and asymmetrical P-Zeros to improve cornering power. The Serie Speciale alone has an F-40-esque front spoiler, and its Kevlar seats lower weight and make you suffer for sport.

Some say a car book criticizing this Ferrari's less-than-supreme performance, ergonomic shortcomings, and lofty pricetag is like *Consumer Reports* disparaging a Rolex watch for lacking the functions of a Timex Ironman. Still, the Serie Speciale's performance only barely tops that of the Chevrolet Camaro Z28, which costs about the same as the sales, luxury, and gas-guzzler taxes on the Ferrari. And a Toyota Supra Turbo will almost certainly whip it around any road-racing circuit. Also, the Acura NSX, like the Ferrari, a mid-engine design, produces near identical performance while being, unlike the Ferrari, comfortable to drive slow and easier to drive fast. (With significant steady-state understeer, the Ferrari is docile at eight tenths, but charge an onramp using a race-pace lower gear and crowd the gas on the exit, and that push instantaneously transforms to big oversteer.) Also unlike the Ferrari, the NSX' spoiler doesn't drag on everything taller than Bott's dots, its shifter doesn't demand a big bicep, and you don't have to rest your chin on the steering-wheel hub to read the speedometer. And it costs half as much. The counter-argument from *tifosi*: "But it's not a Ferrari, is it?" That, we think, means it doesn't have the excitement, the flavor of a Ferrari. And that's sufficient rationale for 100 Americans: The entire run of Serie Speciales will soon be sold.

—Mac DeMere

TECH DATA

Ferrari 348tb Serie Speciale

GENERAL

Make and model	Ferrari 348tb Serie Speciale
Manufacturer/importer	Ferrari North America, Inc., Englewood Cliffs, N.J
Location of final assembly plant	Maranello, Italy
EPA size class	Two-seater
Body style	2-door, 2-passenger
Drivetrain layout	Mid-engine, rear drive
Airbag	None
Base price	$107,300
Price as tested	$111,800
Options included	None
Ancillary charges	Gas guzzler tax, $3000; preparation $350; transportation $1150
Typical market competition	Combination of a Mazda RX-7, GMC Crew Cab Dually, and Lincoln Town Car; 559 haircuts from Cristophe; 1 hour and 21 minutes with Demi Moore

DIMENSIONS

Wheelbase, in./mm	96.5/2450
Track, f/r, in./mm	59.1/64.3/1502/1633
Length, in./mm	170.1/4320
Width, in./mm	74.6/1894
Height, in./mm	46.1/1170
Ground clearance, in./mm	4.8/122
Manufacturer's base curb weight, lb	3232
Weight distribution, f/r, %	40/60
Cargo capacity, cu ft	7.0
Fuel capacity, gal	23.2
Weight/power ratio, lb/hp	10.4

ENGINE

Type	V-8, Liquid cooled, cast aluminum block and heads
Bore x stroke, in./mm	3.35 x 2.95/ 85.0 x 75.0
Displacement, ci/cc	208/3405
Compression ratio	10.4:1
Valve gear	DOHC, 4 valves/cylinder
Fuel/induction system	Multipoint EFI
Horsepower	
hp @ rpm, SAE net	312 @ 7200
Torque	
lb-ft @ rpm, SAE net	229 @ 4000
Horsepower/liter	91.6
Redline, rpm	7500
Recommended fuel	Unleaded premium

DRIVELINE

Transmission type	5-speed manual
Gear ratios	
(1st)	3.21:1
(2nd)	2.11:1
(3rd)	1.46:1
(4th)	1.09:1
(5th)	0.86:1
Axle ratio	3.56:1
Final-drive ratio	3.49:1
Engine rpm,	
60 mph in top gear	2900

CHASSIS

Suspension	
Front	Upper and lower control arms, coil springs, anti-roll bar
Rear	Upper and lower control arms, coil springs, anti-roll bar
Steering	
Type	Rack and pinion
Turns, lock to lock	3.0
Turning circle	39.5
Brakes	
Front, type/dia., in.	Vented discs/11.8
Rear, type/dia., in.	Vented discs/11.8
Anti-lock	Standard
Wheels and tires	
Wheel size, in.	17 x 7.5/17 x 9.0
Wheel type/material	Cast aluminum
Tire size	215/50ZR17/255/45ZR17
Tire mfr. and model	Pirelli P-Zero

INSTRUMENTATION

Instruments	200-mph speedo; 10,000-rpm tach; Oil pressure; coolant temp.; fuel level
Warning lamps	Check engine; oil; battery ABS; slow down; belts

PERFORMANCE AND TEST DATA

Acceleration, sec	
0-30 mph	2.0
0-40 mph	2.8
0-50 mph	4.5
0-60 mph	5.6
0-70 mph	7.6
0-80 mph	9.2
Standing quarter mile	
sec @ mph	14.0 @ 100.0
Braking, ft	
30-0 mph	32
60-0 mph	122
Handling	
Lateral acceleration, g	0.96
Speed through 600-ft slalom, mph	68.8
Speedometer error, mph	
Indicated	Actual
30	29
40	38
50	48
60	57
Interior noise, dB	
Idling in neutral	68
Steady 60 mph in top gear	75

FUEL ECONOMY

EPA, city/hwy., mpg	13/18
Est. range, city/hwy., miles	302/418

the stuff *of* Reims

Forty years ago, Mike Hawthorn won his first grand prix. Phil Llewellin pays homage to his hero by driving to the site of his victory in a Ferrari 348 Spider.

PHOTOGRAPHS BY TIM WREN

Reims circuit on public roads that sweep through cornfields in Champagne country. Faded BP sign a relic of area's former use

THE 1953 FRENCH GRAND PRIX would have been too much for Murray Walker. He would have needed a heart transplant long before the end of what was hailed as the race of the century. The young lion and the old tiger were locked in wheel-to-wheel combat for lap after lap of the 5.2-mile circuit that rockets across the gently undulating cornfields between Reims and Gueux. Ferrari versus Maserati. The tall, blond, debonair Englishman versus the short, swarthy, impassive Argentinian. The new boy versus the maestro. John Michael Hawthorn versus Juan Manuel Fangio.

Beating Fangio by a second, after snatching the lead at the last corner, made the 'Farnham Flyer' the first British driver to win a world championship race. Hawthorn's dramatic victory elevated Reims to a special status in the hearts of British enthusiasts. The circuit was real and heroic in the sense that cars raced on public roads that were closed for the day. One of the best vantage points was the terrace of the Auberge de la Garenne, where corks popped and cutlery clattered within a few feet of the Thillois hairpin.

One of the many evocative photographs in Chris Nixon's book *Mon Ami Mate* reveals how thrilling a grand prix could be when big, front-engined cars raced for almost three hours on skinny tyres. A tablecloth would cover the four leaders – Fangio, Hawthorn, Farina and Villoresi – as they take Thillois in first gear and start accelerating down the long straight that is also the D27.

Buildings that were packed with team managers, mechanics, wives, girlfriends, officials, reporters and thousands of excited fans still flank the road. They are derelict and under threat – the last race was run in 1969 – but retain a powerful magic. It's 40 years since Hawthorn triumphed there, which made it a most appropriate place to go with the Ferrari 348 Spider.

The ragtop was waiting when the ferry from Holland delivered me to Sheerness while sensible citizens were still contemplating their cornflakes. I had been up since five o'clock, and had motored 900 miles the previous day, but medical science has yet to invent a better stimulant than the sight of a Ferrari as red as the Type 500 whose 2.0-litre engine powered Hawthorn into the history books on 5 July 1953. It's easy to be overwhelmed by a Ferrari, because no other marque radiates quite such a potent and seductive aura.

Roots buried deep in motor sport account for a lot of it. Today's 348 Spider may be a far cry from a grand prix car or a 250GTO, but it's impossible not to become a small part of that heritage when you are handed a key with the prancing horse emblem on the

Spider is latest version of 348: V8 engine gives 295bhp, enough to push car to 161mph. Lovely shape the work of Pininfarina: hood design is poor, its cover primitive. Old scoreboard by pits now rusting away (left)

ob, open the driver's door, settle into the seat and run a cockpit check before starting the whooping, wailing engine.

But journalists are paid to be objective, so I was exceptionally conscientious about recording my initial reactions. Styling, for instance, is all about personal tastes. The 348's ability to attract attention is astonishing – you feel like a prize goldfish on display in a bowl – but its lines lack the finesse that made the 308 and 328 look so lithe and sensual.

There's a very real risk of wrecking your jaw on the corner of the windscreen as you drop into the seat. It's trimmed with Connolly leather, but is hard enough to recall the seriously spartan Jaguar D-type that rocketed your ecstatic reporter to Reims and back in 1992.

The 3.4-litre four-cam 32-valve V8's vital statistics are 295 horses at 7200rpm and 234lb ft of torque at 4200rpm. It produces almost as much power per litre as the pure racing engine that did the business for Hawthorn in 1953. But the V8 is a paragon of smooth torque-ing docility as well as a source of elemental pleasure. It starts instantly and will pull sweetly from very low revs if you want to demonstrate flexibility while crawling through traffic.

I know strong, brave men who melt with delight when they see the chromed 'gate' in which the gearlever operates. This baffles me. At best vintage and at worst agricultural, it makes shifting more hit-or-miss than it should be, notably when dog-legging from first to second. Nobody in his right mind treats a cold cogbox with anything other than respect, but engaging the 348's second gear is downright difficult until the transmission has warmed up.

The handbrake is on the right, between driver and door. The warning light didn't go out when it was released, but a simple check – a very gentle start – convinced me it was off duty. Repeating the process after pulling it up proved it was working. Brett Fraser reported the same fault to Maranello Concessionaires after testing the Spider for our sister publication *Carweek*. Later, I was mightily miffed by Maranello's comment that the 348 would have gone faster had Mr Llewellin released the handbrake.

The route to Dover crossed the Isle of Sheppey, whose roads could have been built to test structural integrity. Ferrari's metalwork was impressive, virtually no movements being detected when I wedged my fingertips between sill and door. On this and later evidence I would put the Spider in much the same exalted class as the carved-from-solid Mercedes 500SL.

A police car followed the Ferrari onto the M2, sat on its tail for several minutes, then accelerated and vanished. Visions of being zapped a few miles later dictated a discreet

Four-cam 3.4-litre V8 engine mounted amidships, has smooth low-speed behaviour and storming top-end performance. Cabin sumptuously trimmed, but seats are hard: gated change looks great but disappoints. Instrument graphics let dashboard down. Comprehensive toolkit and Ferrari's own fitted luggage are classy touches

cruising speed all the way to Dover. But it was an interesting exercise, because the V8 revealed a politically correct facet of its character by averaging 24.6mpg while running at an indicated 80mph.

Photographer Tim Wren was waiting at the ferry terminal, where his kit joined my overnight bag up front in the quite practical luggage compartment. The big worry was that the lid is released by a non-lockable lever in the cockpit. Is basic security of no concern to a manufacturer from a country that's as synonymous with theft as it is with supercars? Two other drawbacks had been listed before the *Stena Fantasia* delivered us to Calais. First, the front spoiler is low enough to pose serious problems in commonplace situations. Later in the day, it was impossible to get into the underground car park that makes the Hotel de la Paix a sensible place to stay while visiting Reims with a fancy car.

Second, the instruments' orange markings could have been inspired by the lettering in a stick of Blackpool rock. The dashboard in general is neat enough, apart from cheap-looking switches, but dials with crisp, white-on-black graphics would make it much more attractive and functional.

Getting to Reims in plenty of time for Wren to deploy his Nikons was more important than spurring the Spider along rural roads. The prospect of being fined more than £100 in real money suggested sticking close to the 80mph limit, but at that speed we were being passed by great-grandmothers in pre-war Citroëns. So the right foot eased a little closer to the planet's core and we broke the sound barrier in the sense of reaching the speed at which conversation becomes difficult.

We were running with the hood down, of course. Adopting the *alfresco* mode involves cranking a Heath Robinson handle, because £80,999.99 does not buy push-button convenience. Ferrari, Aston Martin and Rolls-Royce are way behind Mercedes when it comes to dealing with a furled hood. What's needed is a flush-fitting steel panel. What you get, in this case, is a black fabric cover that looks like a huge, pulsating slug. We feared for its safety, because one of the fasteners kept popping, and eventually exiled it to the luggage compartment.

Cruising to Reims in little more than 90 minutes told me next to nothing about the Spider's handling, but a lot about its ride and directional stability. The suspension is firm, but in the calculated and controlled manner that provides sporting sharpness rather than rib-rattling discomfort. You need to keep a firm grip on the wheel, but the kickbacks through the superb rat-and-pigeon steering are cushioned rather than wrist-cracking sharp. Windsocks warned of brisk breezes at several points along the

autoroute, but only once did the Ferrari demand a slight steering correction.

The occasional desire for a little more pace focused attention on the longitudinally mounted V8's storming top-end performance. Fifth is high enough to be quite economical – we averaged 19.6mpg from Calais to Reims – but low enough to be very responsive when you articulate your right ankle. In theory, peak power in fifth corresponds to 170mph.

Colleagues who have driven the Ferrari 348 Spider on test tracks warn that the enormous Bridgestone Expedia ZR tyres – 215/50s in front, 255/45s at the rear – let go in a bowel-watering fraction of a second when the limit of adhesion is reached. I can believe it. My modest experiments on the old circuit, and on the country roads used for part of the return trip, made it abundantly clear that you need to generate a hell of a lot of centrifugal force to get the Spider out of shape. At speeds compatible with having

REIMS 1953

Start of '53 French Grand Prix, with Gonzales and Fangio blasting into lead in A6GCS Maseratis: eventual winner Hawthorn is seventh in this picture. Pits, grandstands still exist

A pensive-looking Mike Hawthorn readies himself for the GP: green windcheater and dark helmet were trademarks. His Ferrari 500 had a 2.0-litre four-cylinder engine giving 190bhp

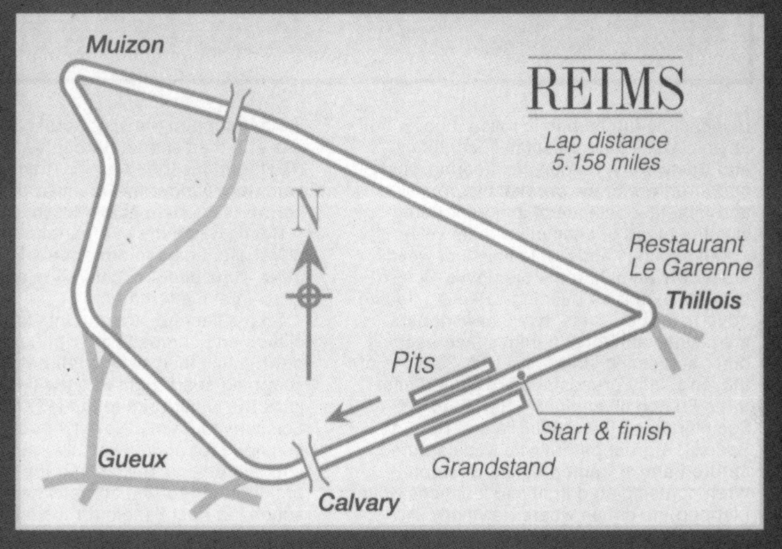

A true road course, this, with the main straight being part of the Reims to Soissons road. Slipstreaming vital for success. Track last used for a GP in 1966, when Jack Brabham won

Stands (far left) named after great French drivers: this one dedicated to Jean-Pierre Wimille. Writer Llewellin never breached high limits of Spider (left), but took heed of warning that car lets go dramatically

responsible fun on public roads it never felt anything less than delectable, confident and trustworthy, reacting to steering wheel and throttle with the predictable precision and beautiful balance of a thoroughbred with the blood of champions in its veins.

Tall saplings and a wilderness of weeds are slowly engulfing the old Reims-Gueux circuit's extensive buildings. Peeling, fading adverts for oils, fuels, tyres, newspapers, magazines and local garages face each other across the start-finish line. Sections of the long, lofty grandstand are named after three French champions – Raymond Sommer, Jean-Pierre Wimille and Robert Benoist, a great patriot who was captured, tortured and strangled by the Gestapo. Wren contemplated light and locations while I lapped the circuit where Hawthorn and Fangio fought their great duel all those years ago. Strong enough to sprint from zero to 60mph in just under six seconds, the 348 topped the slight crest beyond the pits and then snarled through what Hawthorn described as the 'tremendously fast' right-hander that claimed the life of his Ferrari team-mate Luigi Musso in 1958.

Reims was more of a car circuit than a driver circuit, he told readers of *Champion Year*. Top speed, acceleration and brakes were what mattered here.

So it's flat and straight until the road climbs and curves to the right, swoops down a hill, then swings left before snapping sharp right into the hairpin that joins the minor road to the N31 from Soissons to Reims. Now it's boot-to-bulkhead for about two miles before diving and climbing to Thillois and the hard-braking, fifth-to-first corner where Hawthorn squeezed past Fangio for the last time, within sight of the finishing line. Gregor Grant's report in *Autosport* – 'The excitement was indescribable' – tells how they passed the pits wheel-to-wheel on 10 of the 30 laps that embraced their duel.

Hawthorn became the first British world champion by finishing second to Moss in the Moroccan Grand Prix, 1958's last race. He won only one grand prix that year, but his consistency and the Ferrari's reliability were enough to thwart Moss and Vanwall.

'Snowball' retired at the end of the season that had been blackened by the death of his great friend, Peter 'Mon Ami Mate' Collins, but was killed on 22 January 1959. His modified 3.4-litre Jaguar crashed near Guildford, Surrey, during a dice with a Mercedes 300SL driven by Rob Walker. The swashbuckling driver was just 29.

Hawthorn's grand prix career spanned seven seasons, but produced only three world championship wins from 45 starts. He cannot be rated among the truly great champions, probably because his attitude was essentially amateur, but the victory

Remains of old circuit stand out like ancient ruins in rolling fields of Reims area. Safety measures were minimal

over Fangio proved him to be a driver of exceptional maturity, courage and tenacity.

I never met Hawthorn, and saw him race only once, but he was my kind of hero, the sort of pipe-smoking, tweed-capped cavalier who would have piloted a Spitfire had he been born a decade earlier. He was a demi-god on the track, where his kit included a bow tie, a green windcheater and battered brogues, but there was a very human side to his character. It was easy to imagine having a tremendous night out with Mike Hawthorn. He ate fish and chips, loved drinking beer and playing darts in pubs, and paused to collect a therapeutic pint during his slowing-down lap after finishing second in the '58 British GP. Earlier that season he had jumped into a bath fully dressed, then stayed up until six o'clock in the morning, playing poker with three Florida cops after his Ferrari 250TR failed to finish the sports car race at Sebring.

Countless young ladies were captivated by what Chris Nixon calls 'a bowl-'em-into-bed smile' worthy of Jack Nicholson. We recalled this while relaxing over an excellent meal in what is now the Restaurant la Garenne. Hawthorn had dined there on the night of his great victory, sharing a table with another British driver and two local beauties, one of whom gave birth to a son exactly nine months later. She called him Mike.

They had met at a Reims Tennis Club cocktail party, as described by Nixon in *Mon Ami Mate*: 'Mike was adept at inviting himself to parties. If he saw some pretty girls through a window and heard the sound of clinking glasses, he was quite capable of walking in off the street and joining in the fun, his tremendous charm and that dazzling smile obviating the need for the formality of an invitation.'

Three names – Ferrari, Lotus and Hill – were just about the only links with Hawthorn's era at this year's French Grand Prix. I had one of the best seats, but was sleeping when Schumacher's passing of Senna constituted the afternoon's only split-second of genuine excitement. In contrast, I recalled how Hawthorn described 1953's battle in *Challenge Me The Race*: 'Positions were changing several times a lap. I had the lead and then Villoresi came past, and sometimes we would be hurtling along three abreast, at 160mph, down an ordinary French main road. We would go screaming down the straight side-by-side absolutely flat out, grinning at each other, with me crouching down in the cockpit, trying to save every ounce of wind resistance. We were only inches apart and I could clearly see the rev counter in Fangio's cockpit.'

Picture that next time you're trying to stay awake while watching Bernie Ecclestone's androids on the TV.

FIRST LOOK With the factory returning to sports-car racing in a serious way, U.S. privateers are looking once more to Ferrari.

Story by Rich Taylor, photos by Jean Constantine.

Competizione!

Think back to IMSA's 24 Hours of Daytona. What do you remember? Maybe not the GTS Nissan that won, and probably not the WSC prototypes that didn't. The car you do remember is the red, white & blue Ferrari 348 GT Competizione that got all the attention and all the publicity, even though it only finished 8th in GT and 16th overall. Heck: There's just a charisma about Ferraris that fascinates people.

But maybe there's something more to it as well. The charismatic Competizione that appeared at Daytona was not the product of a lavish factory racing effort—in fact just the opposite, to hear the principals tell it.

The car is privately owned by Tom and Steve Shelton, Ferrari dealers and amateur racers from Fort Lauderdale, Florida, and it was driven at Daytona by the two of them plus former Formula One and Indycar star Didier Theys. The Daytona effort was underwritten by Ferrari enthusiast Arthur Coia—his son, Art Coia, Jr., did about five laps at Daytona just to be able to say he got in the car during the race, but essentially the Coias supported the team because they simply love Ferraris.

Still, the GT Competizione probably got more than its share of attention at Daytona for a very simple reason that has little to do with abstract attraction. Privateer effort or not, it was symbolic of Ferrari's long-awaited return to sports-car racing as a whole.

Rumors alluded to the Competizione as a stand-in for the yet-to-be-introduced 333SP WSC racer. Ferrari is also actively pursuing its 348 Challenge series (see SCI March 1994) in America and abroad, and a passel of 348 Competiziones in GT racing would certainly have shades of a N.A.R.T.-esque effort, modest privateer budgets or no.

Even a modestly budgeted IMSA effort is not cheap, of course. Tom Shelton admits that the car itself, not counting what it cost to run the 24-hour race, came in at $259,300.

He refused to tell me his expenses, but I know from my own 24-hour racing that the team's budget must have been in the $100,000 range for that one weekend, especially since they didn't have a tire deal and had to pay real money for the fifty Goodyear racing skins they consumed during the race. Since then, the Sheltons have been unable to find spon-

Even a modestly budgeted IMSA effort is not cheap, of course. Tom Shelton admits that the car itself, not counting what it cost to run the 24-hour race, came in at $259,300.

sorship to continue the rest of the IMSA season, so the car sits today in their Fort Lauderdale, Florida, dealership, spiffed up and ready to race. It's a one-off piece of racing history, and a pretty interesting technical exercise as well—one in building a racecar out of a street car, in a very short period of time and on a relatively limited budget.

How did it all come about? Well, according to Tom Shelton, "Dr. Buitoni, the president of Ferrari North America, called me one day last fall and asked if I'd like to buy a 348 GT that was 'similar' to the car in which Oscar Laurrari had won the Italian GT Championship. Dr. Buitoni faxed us some specifications, and my brother and I ordered the car."

As the 348 arrived from Ferrari racing specialist Michelotto, it had been given carbon fiber bumpers front and rear, carbon fiber doors and a hood and decklid of thin aluminum with a carbon fiber

Why Daytona?

➤ *The 348 GT Competizione garnered attention at Daytona this year for one more reason that Rich doesn't mention; there's a lot of emotion and history tied up between Ferrari and this track. Just look at the car that more than one old crock has declared the last "true" Ferrari: the 365GTB/4 Daytona.*

➤ *Why that name? The accepted tale is that Enzo Ferrari was so pleased with his team's 1-2-3 sweep of the 1967 Daytona 24-hour—Ferrari had been positively pantsed by Ford in '65 and '66—that he determined to name his next berlinetta after the famous Florida track.*

➤ *Word leaked out to the press, though, so the ever-mercurial Drake stamped his feet and declared he'd simply call the car 365GTB/4 instead, using Ferrari's standard system of nomenclature. No beady-eyed journalists were going to get the best of him.*

➤ *When the new GT arrived, however, everybody was already so enamored with the name Daytona that it stuck, and remains with us to this day—joining the Gullwing, TR3A and many others that are universally known by names their factories never endorsed.*

➤ *Frankly, I've always found this particular story just a little too pat myself, but I repeat it here because it's one of the few tales of any kind that Ferraristi generally agree on!*

➤ *Regardless, like the 348 Competizione, the Daytona became a serious racing car only through privateer—distributor, actually—efforts. Sixteen alloy-bodied Daytonas were constructed by Scaglietti in all, an early one for the 1969 Le Mans 24-hour (DNS) and three later batches of five competition cars each.*

➤ *Though they got off to a shaky start, the Daytona competition cars went on to be venerable workhorses of 1970s GT racing—from Sebring, 1971, right up to an emotional second place in the Daytona 24-hour of 1979.* —*George Stradlater*

PHOTO CONSTANTINE

stiffener underneath. The rear window was Lexan, while the interior had been stripped and fitted with an OMP racing seat covered in gray cloth. Michelotto had also installed an FIA-legal rollcage.

Michelotto's racing engine was rated at 360 bhp, compared to the 300 horses of a stock 348. It had different fuel mapping in the computer, but stock cams, stock pistons and mostly stock internals. In addition, Michelotto had fit the car with F40 halfshafts, 15-inch cross-drilled Brembo discs and 18x8-inch wheels in the front, 18x10-inch wheels at the rear.

It had Koni racing shocks and progressive-rate springs. The frame had been strengthened by welding and bolting stiff gussets around the engine compartment. There was a bleeder for the cooling system and NACA ducts in the airdam for cooling the front brakes. Michelotto placed a carbon fiber undertray below the engine compartment to reduce buffeting at high speeds and fit it with NACA ducts to lead air to the rear brakes. The Shelton's new racecar weighed 2630 pounds when it arrived, compared to 3050 pounds for the stock Euro-spec Ferrari 348.

Typically Italian (or typically race program), when they finally got the car through Customs there were only two weeks left to prepare it for Daytona. The Sheltons turned their service department upside down and hired Ferrari expert Mac MacBride, of European Exclusive, to do most of the fabrication. In a very real sense, the "race ready" 348 GT built by Michelotto was merely a place to start—an assemblage of parts that could be turned into an IMSA car.

Shelton's mechanics installed four air jacks for quick tire changes—an absolute necessity for endurance racing—under the four corners of the chassis. They changed the brake rotors to ones that hadn't been cross-drilled so there'd be less chance of cracking. They redesigned the front brake ducting to make airflow more direct—straight through the front bumper rather than by NACA ducts in the airdam. They also made ducting to get cool air to the overworked alternator—running at high speeds with halogen headlights and driving lights, the electrical system is constantly being strained. As it came from Michelotto the car had standard flip-up headlights, but Shelton's team fabricated a faired-in headlight system with 100-watt halogens and driving lights for extra illumination.

Michelotto provided an intercooler for the transaxle, heavy-duty shift cables, beefier shift fingers and stronger transmission gears. They also came up with center-lock BBS wheels, 18x9-inch front and 18x11.5-inch rear, and the bolt-on conversion kit to mount them on the earlier spindles.

The Sheltons added a 22-gallon fuel cell, a 4-gallon surge tank and six fuel pumps. (Two pumps supply fuel to the surge tank and two move fuel from the surge tank to the engine. Another two pumps can be switched on by the driver to empty the last two gallons out of the sump of the surge tank.)

They also fabricated sidebars for the rollcage and changed the interior ventilation system to get more air to the hard-working driver. The engine was fit with IMSA-required restrictors—essentially gutted catalytic converters—which reduced horsepower to about 350 bhp at 8000 rpm. When presented to IMSA's technical inspectors, the car weighed 2750 pounds, or 320 pounds more than the minimum weight in the GT class.

Because they were so late getting the car from Michelotto, the first time anyone drove it was at Daytona, Thursday night before the race, during night practice. The team immediately learned they didn't have the proper springs; it handled beautifully in the infield but would squash right down when it hit the banking.

Daytona is a very difficult track to set up for properly. Because of the G-loads on the banking, you have to compromise

all your suspension settings. The Sheltons ran 2200-pound front springs and 26.5-inch-tall Goodyear tires instead of the 1600-pound springs and 25-inch-tall Pirellis that Michelotto had designed the car around. Ultimately, changing the suspension geometry like this put a lot of load into the tires and caused severe wear: All the suspension compliance was essentially just sidewall flex.

During the race, the Ferrari cut two tires and had two more flats on the overstressed right front. They lost about 45 minutes in all, between getting the car slowly back to the pits, changing the tires and getting back out into the race. They also had a problem with the rear brake calipers hitting the rotor "hat" and locking the rear brakes.

As Tom Shelton says, "You may have seen me spinning on ESPN. They said it was Steve, but it was actually me. The rear brakes had locked and there was nothing I could do. Without the brake and tire problems—we went through 50 tires in 24 hours—we would have been in the top ten overall. We had zero mechanical problems; *zilch.* I think it could go run another 24-hour race right now."

The Ferrari 348 GT is a racing car. It's hot and noisy and stiff, but very quick and predictable, this according to Tom Shelton. And also very exciting to drive: It was timed at 172 mph on the banking at Daytona, but as Shelton explains, "We were way overgeared once we switched to the taller Goodyear tires. We were only seeing 7000 rpm, but the car is designed for 8500."

The 348 GT was significantly slower than the class-winning Porsche RSRs at Daytona. Didier Theys lapped at 2:05 in practice and ran 2:08s in the race; Steve and Tom turned laps of 2:09 or 2:10. The RSR Porsches, by comparison, were running 2:03s in race trim.

"This car needs some development to be competitive in IMSA's GT class," admits Tom Shelton. "It handles well, but it's overweight." On the other hand, this particular 348 GT has electric windows and air jacks for endurance racing. Some of this weight could be easily unbolted. Plus, Michelotto has already developed carbon fiber front fenders for the car and plexiglass windows.

As Tom Shelton says, "I don't think you could get it down to IMSA's minimum weight of 2430 pounds without a whole lot of expensive titanium pieces, but it could easily become weight competitive with the Porsche RSRs. And we've already ordered a 400 bhp engine from Michelotto."

What are they going to do with this car? Tom Shelton is looking for someone to sponsor his team in IMSA, or someone who'd like to buy the car and race it, with or without his help. His dealership has also ordered two more 348 GT Competiziones from Michelotto, each costing about 50% more than a stock 348. But since there won't be more than six or eight built, total, that ought to be a pretty good investment for a die-hard Ferrari racer—or collector.

So, ah...looking for charisma? ○

I don't think you could get it down to IMSA's minimum weight of 2430 pounds without a whole lot of expensive titanium pieces, but it could easily become weight competitive with the Porsche RSRs.

Specifications

Ferrari 348 GT Competizione (Shelton)

➤ General
Vehicle type: mid-engine, rear-wheel-drive coupe
Structure: sheet- and tubular-steel frame with steel, aluminum and carbon fiber body panels
Market as tested: competition
Base MSRP: $259,300

➤ Engine
Type: longitudinally-mounted V8, aluminum block and heads
Displacement (cc): 3405
Bore x stroke (mm): 85 x 75
Compression ratio: 10.4:1
Horsepower (bhp): 350
Intake system: MPFI
Valvetrain: two overhead cams per bank, four valves per cylinder
Transmission type: 5-speed manual

➤ Dimensions
Curb weight (lbs.): 2630
Wheelbase (in.): 86.5
Length (in.): 166.7
Width (in.): 74.6
Fuel capacity (US gal.): 22 (+4-gallon surge tank)

➤ Suspension, brakes, steering
Suspension, front: double wishbones with coil springs and antiroll bar
Suspension, rear: double wishbones with coil springs and antiroll bar
Steering type: rack and pinion
Wheels, f/r: 18x9/18x11.5 (BBS)
Brakes, f/r: 15-inch vented disc/15-inch vented disc

348 SPIDER

Sensual necessity

BY RAY THURSBY

PHOTOS BY BRIAN BLADES

Of all the cars on these pages, there is one, and in my view, only one, that you *need* to drive and, circumstances willing, own. You could live happily without ever possessing a Japanese supercoupe, or a Detroit ponycar, or a Teutonic techno-marvel, but to deny yourself the pleasure of a relationship with a Ferrari 348 spider means you'll never know what a sports car—a *real* sports car—is all about.

Why the 348? Is it the best-looking of automobiles? Close, maybe, but no. The fastest? Surely not. Swiftest around slalom and skidpad, then? No. Most comfortable, best-built or most efficient? Naaah.Take it from me: It's the Mystique Thing that makes the spider special.

We beady-eyed motoring writers are supposed to be immune to that sort of influence. Seen it, been there, done that, you know. The spider is just a car: It makes contact with the ground via four smallish patches of rubber, has valves, pistons, gears, hoses and wires, and is made from the same kinds of raw materials that go into other cars. Uses fuel out of an ordinary pump and gets dirty when exposed to the elements, as does any car you care to name.

Somewhere along the line, though, the spider has been endowed with a soul.

Actually, there are a trio of 348s, and all have souls. There's a coupe (tb), a targa (ts), and the spider as tested, in ascending price order. The comments made here apply equally to all of them, except for the open-air bits.

First impressions set the stage. Not only is the 348 instantly identifiable as a Ferrari, but it is also a lovely piece of work. The body seems barely able to conceal the wonders below, and does nothing to hide the need for large quantities of cooling air; likewise, the spatial requirements of wheels, powerplant and passengers are made obvious without use of styling tricks or faddish add-ons. To my eyes—and, as was made clear during my time with the car, to many others—the 348's exterior (a Pininfarina design) is a success, even if it doesn't quite match the voluptuous grace of its Dino 246 grandparent.

Don't stop there. Push the button on the spider's door and slide into the cockpit. Initially, it's a bit of a chore, since you have to worm your way past the silly reverse mouse belt that slides along the bottom of the door sill and is anchored behind your right shoulder. Practically everything save the floor is swathed in leather; smells good, feels good. The seats are firm, snug, and just right. The driving position is configured for the Standard Italian Bus Driver, he of the permanent slouch, long arms and short legs. Never mind; do as the Italians do and grasp the wheel rim by its lower half. It's more comfortable than it looks.

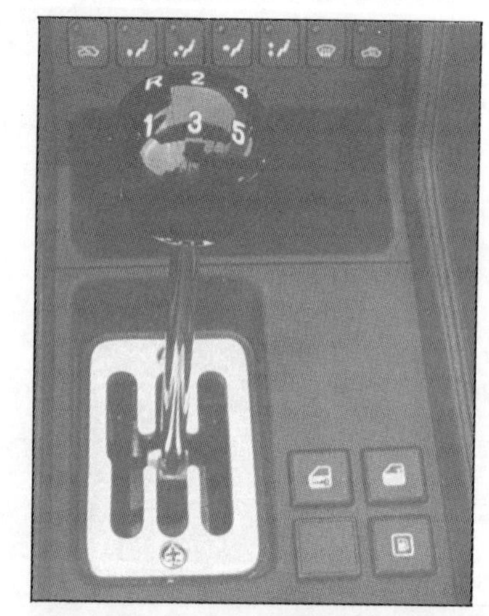

It's the Mystique Thing that makes the spider special.

By this point, you will have already encountered a few eccentricities. Outside, the 348's nosepiece sports a fake air inlet (which would only cool the luggage bin were it real). Inside, two vital gauges are in the center console, out of range of quick glances. And then there's the top. Handsome (and wind- and watertight) when raised, it can be lowered quickly, with a few Italian comic-opera touches: A clever single latch secures it at the header, but a curious over-center crank lowers it into its final resting place. After, that is, you've opened both doors and folded both seatbacks forward. Practice in private before you set out to amaze your friends.

All quirkiness evaporates when you turn the ignition key. Though the 348 has a V-8 engine (and a very robust one it is, churning out 310 bhp from 3.4 liters, thanks to four camshafts, 32 valves, and a state-of-the-art Bosch Motronic injection system), it sounds as much like a domestic bent-eight as Enrico Caruso sounds like Bob Dylan. In turn, it snarls, bellows, shrieks, all the way to its 7500-rpm redline. You will never miss the radio that isn't standard equipment. If raucous, the spider's all-alloy powerplant is otherwise civilized, being equally at ease idling around town or running at maximum out somewhere beyond earshot of spoilsport authorities.

Once underway, the 348 sends mixed messages to its driver, which may account for the variety of opinions expressed by writers of past road tests. The rack and pinion steering is on the heavy side (no power assist, plus wide tires), the brakes ditto (ABS is standard, but you need the leg strength of a Sumo wrestler to engage it) and the shifter is sluggish (and noisy as it contacts the gate).

All this remains true, and perhaps annoying, until you learn the secret to taming the 348: When it resists, *push harder.* That's right, go ahead; it won't break. After a few miles, it all begins to work, and you and spider begin to dance to the same rhythm. I don't say it's easy, mind you, but few really good experiences are effortless.

At that point fast driving becomes less of a calf-roping competition and more of a horse race. The spider remains graceful right up to its limits, ultimately exchanging its neutral stance for easily controlled oversteer, and the driver is always in control, always knows exactly what's happening. Shifts, turns, accelerations and decelerations are all delivered promptly and precisely as ordered. Keep the revs up—between 4000 and 7000 rpm is best—and the spider rockets away from corners. I can't imagine anyone who has had the pleasure of driving a 348 over a challenging road wanting to change a single part or system.

Said systems are made up, as always in a Ferrari, of simple and highly developed components. The chassis is a welded tubular structure, each alloy wheel is backed by a disc brake and a basic upper and lower A-arm/coil spring suspension unit, and the engine drives through a 5-speed gearbox set transversely in the frame in the interests of good balance. The transmission's ratios are perfectly spaced for aggressive driving.

Along with these delights come a few more quirks. While exterior panels have surface and paint finishes that rival the best custom work, other details will be the despair of the concours-bound. Visible frame tubes and their environs are covered by a protective but hardly glamorous layer of flat-black paint. Save for red crackle paint on the cam covers and black crackle finish on the air-cleaner box, the engine's castings are as they left the foundry. Everything is assembled properly and is arranged for relatively easy service, but those whose previous Ferrari experience came on the lawns of Pebble Beach will be in for a rude awakening.

We've come full circle now, right back to the Mystique Thing. The 348 is fast, as the data panel shows, and feels faster, but isn't the speed king of its class. It is beautifully detailed (at least where it counts), but not to the injection-molded level of top-line Japanese sportsters. It's more fun to drive, fast or slow, than just about any other car you'd care to name, but in the process strains muscles you didn't know you had. It does more to enhance its driver's image than plastic surgery, but costs more than most houses (somewhere around $150,000 by the time the dealer, Ferrari, and various state, local and national agencies get their cuts).

What you get for the money is a Ferrari. To say that driving this (relatively) mass-produced Ferrari gives one a kinship with heroic types like Ferrari pilots Fangio the Elder, Phil Hill and Gilles Villeneuve may be laying it on a bit thick, but you can't help getting a touch of that feeling from behind the spider's wheel. And you can't avoid the feeling that every drive, short or long, is an *event*.

That's what Mystiques—and the 348 spider—are all about.

F E R R A R I

348 SPIDER

PRICE

List price, all POE	$121,900
Price as tested	$132,155

Price as tested includes std equip. (air cond, leather interior, ABS, pwr windows, mirrors & door locks), gas guzzler tax ($3000), dest charge ($1150), luxury tax ($9105).

ENGINE

Type	dohc 32-valve V-8
Displacement	3405 cc
Bore x stroke	85.0 x 75.0 mm
Compression ratio	10.4:1
Horsepower, SAE net	310 bhp @ 7200 rpm
Torque	229 lb-ft @ 4000 rpm
Maximum engine speed	7500 rpm
Fuel injection	Bosch Motronic M2.7
Fuel requirement	premium unleaded

GENERAL DATA

Curb weight	3252 lb
Test weight	3415 lb
Weight distribution, f/r, %	est 40/60
Wheelbase	96.5 in.
Track, f/r	59.2/63.1 in.
Length	166.7 in.
Width	74.6 in.
Height	46.1 in.
Trunk space	6.8 cu ft

CHASSIS & BODY

Layout	mid-engine/rear drive
Body/frame	steel & aluminum/steel skeleton & tubular steel
Brakes, f/r	11.8-in. vented discs/12.0-in. vented discs, vacuum assist, ABS
Wheels	cast-alloy, 7.5J x 17 f, 9.0J x 17 r
Tires	Pirelli P-Zero, 215/50ZR-17 f, 255/45ZR-17 r
Steering	rack & pinion
Turns, lock to lock	3.25
Suspension, f/r	upper & lower A-arms, coil springs, tube shocks, anti-roll bar/upper & lower A-arms, coil springs, tube shocks, anti-roll bar

DRIVETRAIN

Transmission .... 5-sp manual

Gear	Ratio	Overall Ratio	(Rpm) Mph
1st	3.21:1	13.93:1	(7700) 41
2nd	2.11:1	9.16:1	(7700) 63
3rd	1.46:1	6.34:1	(7700) 90
4th	1.09:1	4.73:1	(7700) 121
5th	0.84:1	3.65:1	(7700) 157

Final-drive ratio	4.34:1
Engine rpm @ 60 mph, top gear	2900 rpm

ACCELERATION

Time to speed	seconds
0-30 mph	2.1
0-40 mph	3.0
0-50 mph	4.5
0-60 mph	5.6
0-70 mph	7.5
0-80 mph	9.1
0-90 mph	11.7
0-100 mph	14.0
Time to distance	
0-100 ft	2.9
0-500 ft	7.7
0-1320 ft (1/4 mile)	14.1 sec @ 101.0 mph

BRAKING

Minimum stopping distance	
From 60 mph	140 ft
From 80 mph	245 ft
Control	excellent
Overall brake rating	excellent

HANDLING

Lateral accel (200-ft skidpad)	0.91g
Speed thru 700-ft slalom	62.8 mph

FUEL ECONOMY

Normal driving	est 15 mpg
Fuel economy (EPA city/highway)	13/19 mpg
Fuel capacity	23.2 gal.

est estimated, na means information not available